Benzene Carcinogenicity

Editor

Muzaffer Aksoy

Professor of Medicine
The Scientific and Technical Research
Council of Turkey
Research Institute for Basic Sciences
Department of Biology
Gebze, Kocaeli, Turkey
and
Department of Internal Medicine
Section of Hematology
Istanbul Medical School, Çapa
Istanbul, Turkey

CRC Press
Taylor & Francis Group
Boca Raton London New York

CRC Press is an imprint of the
Taylor & Francis Group, an **informa** business

First published 1988 by CRC
Press Taylor & Francis Group
6000 Broken Sound Parkway NW, Suite
300 Boca Raton, FL 33487-2742

Reissued 2018 by CRC Press

© 1988 by Taylor & Francis
CRC Press is an imprint of Taylor & Francis Group, an Informa business

No claim to original U.S. Government works

A Library of Congress record exists under LC control number: 87008048

Publisher's Note
The publisher has gone to great lengths to ensure the quality of this reprint but points out that some imperfections in the original copies may be apparent.

Disclaimer
The publisher has made every effort to trace copyright holders and welcomes correspondence from those they have been unable to contact.

ISBN 13: 978-1-138-50619-0 (hbk)
ISBN 13: 978-1-138-55769-7 (pbk)
ISBN 13: 978-1-315-14997-4 (ebk)

Visit the Taylor & Francis Web site at http://www.taylorandfrancis.com and the CRC Press Web site at http://www.crcpress.com

THE EDITOR

Muzaffer Aksoy, M.D., is Professor of Medicine, Section of Hematology, Department of Internal Medicine, at Istanbul Medical School, Çapa, Instanbul, Turkey and a member of the medical research group TÜBITAK (Scientific and Technical Research Council of Turkey), Unit of Hemolytic Anemias and Chemical Hematotoxic Disorders. He is also a Fellow of Collegium Ramazzini, Modena, Italy.

After Graduating from the Medical School of Istanbul University in 1940, he served for 4 years in the Turkish Navy and Army as a reserve medical officer. In 1959, he became Assistant Professor of Medicine and became Professor of Medicine in 1966.

Since 1960, he has worked in the Internal Clinic of the Medical Faculty of Istanbul University. He has delivered several lectures throughout the U.S. and Europe and participated in numerous symposiums world-wide as a contributor. He has been a member of the editorial committee of the *New Istanbul Contribution to Clinical Science* since 1962, a member of the editorial board for *Hemoglobin* magazine since 1976, and in the past 20 years has published numerous papers on the subject of hematology and its associated disorders.

Muzaffer Aksoy is currently active in research concerning blood disorders, such as leukemia, anemia, thalassemia, and related, especially those disorders which are a result of chemical/carcinogen exposure.

CONTRIBUTORS

Muzaffer Aksoy
Professor of Medicine
Department of Internal Medicine
Section of Hematology
Istanbul Medical School, Çapa
Istanbul, Turkey

Keith R. Cooper
Associate Professor
Department of Biochemistry and
Microbiology
Rutgers-The State University
New Brunswick, New Jersey

Masayuki Ikeda
Professor
Department of Environmental Health
Tohoku University School of Medicine
Sendai, Japan

Toshihide Okuno
Associate Director
Department of Chemicals Survey
Environmental Science Institute of
Hyogo Prefecture
Kobe, Japan

Carroll A. Snyder
Research Professor
Department of Environmental Medicine
New York University
Tuxedo, New York

Robert Snyder
Chairman
Department of Pharmacology and Toxicology
Rutgers-The State University
New Brunswick, New Jersey

TABLE OF CONTENTS

Chapter 1

PROPERTIES, OCCURRENCES, AND EMISSIONS

Masayuki Ikeda

Benzene (C_6H_6, molecular weight = 78.1; Chemical Abstract Service Registration Number 71-43-2) is clear, colorless, flammable liquid at room temperature, and has an aromatic odor. The specific gravity at 15°C, referred to water at 4°C (d_4^{15}), is 0.8787. It boils at 80.1°C (at 760 mmHg) and solidifies at 5.5°C. It is soluble in 1430 parts of water, and miscible with alcohol, chloroform, ether, acetone, and other organic solvents as well as oils.[1] The n-octanol/water partition coefficient (Po/w) is 135.[2] The flash point is −12 to −10°C, and the ignition temperature is 490°C. The vapor density is 2.7 (air = 1.0) and the explosive range for vapor is 1.4 to 7.1% by volume in air.[3] The liquid is volatile with vapor pressure of 100 mmHg at 26.1°C. Benzene is readily biodegradable in the presence of activated sludge;[4] 1 ppm benzene is equivalent to 3.19 mg/m³.

Although benzene had been commercially recovered from both coal and petroleum sources in former days (benzene is a natural constituent of crude oil[5]), the current production of benzene is almost exclusively based on petroleum. For example, about 91% of benzene production in Japan in 1984[6] and 92% of that in the U.S. in 1978[7] were from petroleum. It is primarily used as a raw material for production of various chemicals including cyclohexane, cumene, ethylbenzene, styrene, nitrobenzene, maleic anhydride, chlorinated benzenes, etc.[8] Its application as an organic solvent is very limited. For instance, as early as 1967 in Japan, the amount of benzene used as solvent was already less than 1% of the produced amount.[9] A nationwide survey in Japan on solvent-containing products such as paints, thinners, inks, and adhesives disclosed that benzene was seldomly detected among over 1000 samples collected from factories.[10,11]

Automobile gasoline contains benzene at various levels. The reported figures were, as an average, 0.8% in the U.S.,[12] 1.4% in Japan,[13] and 5% in Europe.[5] Although this might also be the case of wide boiling point range industrial gasoline used as solvent in factories, it should be added that no benzene was found in some high boiling point industrial gasoline products (e.g., Japanese Industrial Standard K 2202 Category No. 5 industrial gasoline or so-called dry cleaning solvent with a boiling point range of 150 to 210°C).[14] Benzene in automobile gasoline can be the sources of direct exposure for gasoline station attendants,[15] bulk handling operators,[16] or even drivers at self-service stations.[7] Benzene comprises about 2.2% of total hydrocarbon emissions from a gasoline engine, or about 4% of automobile exhaust.[12]

Benzene levels for ambient air in major cities in the 1970s ranged from a low of 0.3 μg/m³ (0.1 ppb) in Praha, to a high of 573 μg/m³ (179 ppb) in London,[7] in contrast to the rural background level of 0.3 to 54 μg/m³ (0.1 to 17 ppb).[5] Dramatic reduction appears to have taken place in the benzene levels in occupational settings. A level of as high as 3000 mg/m³ (over 1000 ppm) was reported in the 1930s,[3] while recent studies reported levels of less than 30 mg/m³ (less than 10 ppm) in most cases.[17-19] Annual emissions of benzene into the air in the U.S. is summarized as from components (e.g., production, vending, etc.) of gasoline (40 to 80; unit, 10³ tons/year), production of other chemicals (44 to 56), indirect production (e.g., coke oven, oil spills, etc.) of benzene (23 to 79), production of benzene from petroleum (1.8 to 7.3), and solvents and miscellaneous use (1.5).[20]

Analyses of water media in the general environment also has disclosed presence of benzene. For instance, Colenutt and Thorburn[21] detected benzene in lake water (6.5 to 8.9 μg/ℓ), stream water (18.6), river water (6.8), and rainwater (87.2) as well, in the United Kingdom.

The U.S. Environmental Protection Agency,[12] reported the presence of 0.1 to 0.3 μg of benzene per liter of drinking water and over 100 μg/ℓ of groundwater. Sources of benzene emission to water in the U.S. are indirect production of benzene (0.2 to 11 × 10³ tons/year), solvent and miscellaneous use (1.5), production of other chemicals (1.0), and production of benzene from petroleum (0.6).[20]

For environmental standards, occupational exposure limits, and other regulation-related values, see Chapter 5.

REFERENCES

1. **Windholz, M., Budavari, S., Stroumtsos, L. Y., and Fertig, M. N., Eds.,** *The Merck Index,* 9th ed., Merck, Rahway, N.J., 1976, 138.
2. **Kenaga, E. E. and Corning, C. A. I.,** The 3rd Symp. Aquatic Toxicity, New Orleans, October 17, 1978.
3. National Institute for Occupational Safety and Health, Criteria for a Recommended Standard — Occupational Exposure to Benzene, HEW Publ. No. NIOSH 74-137, Washington, D.C., 1974, 22, 23, and 121.
4. Ministry of International Trade and Industry, Japan, *MITI Gazette,* December issue, 1979 (Japanese).
5. **Brief, R. S., Lynch, J., Bernath, T., and Scala, R. A.,** Benzene in the workplace, *Am. Ind. Hyg. Assoc. J.,* 41, 616, 1980.
6. Ministry of International Trade and Industry, Japan, Ed., *Yearbook of Chemical Industries Statistics — 1984,* Tsusan Tokei Chosa Kai, Tokyo, 1984, 71.
7. International Agency for Research on Cancer, *IARC Monographs on the Evaluation of the Carcinogenic Risk of Chemicals on Humans,* Vol. 29, Lyon, France, 1982, 93.
8. **Kagaku Kogyo Nippo, Ed.,** *9285 Commercial Chemicals,* Kagaku Kogyo Nippo Sha, Tokyo, 1985, 577 (Japanese).
9. Ministry of International Trade and Industry, Japan, *Yearbook of Chemical Industries Statistics — 1967,* Tsusan Tokei Chosa Kai, Tokyo, 1967, 70.
10. **Inoue, T., Takeuchi, Y., Hisanaga, N., Ono, Y., Iwata, M., Ogata, M., Saito, K., Sakurai, H., Hara, I., Matsushita, T., and Ikeda, M.,** A nationwide survey on organic solvent components in various solvent products. I. Homogeneous products such as thinners, degreasers and reagents, *Ind. Health,* 21, 175, 1983.
11. **Kumai, M., Koizumi, A., Saito, K., Sakurai, H., Inoue, T., Takeuchi, Y., Hara, I., Ogata, M., Matsushita, T., and Ikeda, M.,** A nationwide survey on organic solvent components in various solvent products. II. Heterogeneous products such as paints, inks and adhesives, *Ind. Health,* 21, 185, 1983.
12. U.S. Environmental Protection Agency, Ambient Water Quality Criteria for Benzene, EPA-440/5-80-018, Washington, D.C., C1-8, C16-35, and C67-100.
13. **Ikeda, M., Kumai, M., Watanabe, T., and Fujita, H.,** Aromatic and other contents in automobile gasoline in Japan, *Ind. Health,* 22, 235, 1984.
14. **Ikeda, M. and Kasahara, M.,** n-Hexane and benzene contents in gasoline for industrial purpose, *Ind. Health,* 24, 63, 1986.
15. **Pandya, K. P., Rao, G. S., Dhasmana, A., and Zaidi, S. H.,** Occupational exposure of petroleum pump workers, *Ann. Occup. Hyg.,* 18, 363, 1975.
16. **Phillips, C. F. and Jones, R. K.,** Gasoline vapor exposure during bulk handling operations, *Am. Ind. Hyg. Assoc. J.,* 39, 118, 1978.
17. **Markel, H. L., Jr. and Elesh, W. E.,** Health Hazard Evaluation Determination Report No. 78-57-579, Amoco Texas Refining Co., Texas City, Tex., National Institute for Occupational Safety and Health, Cincinnati, Ohio, 1979; as cited in **IARC Monographs,** Vol. 29, 1982.
18. **McQuilkin, S. D.,** Health Hazard Evaluation Determination Report No. HE78-102-677, Continental Columbus Corporation, Columbus, Wis., National Institute for Occupational Safety and Health, Cincinnati, Ohio, 1980; as cited in **IARC Monographs,** Vol. 29, 1982.
19. **Hancock, D. C., Moffitt, A. E., Jr., and Hay, E. B.,** Hematological findings among workers exposed to benzene at a coke oven by-product recovery facility, *Arch. Environ. Health,* 39, 414, 1984.
20. JRB Associates, inc., Materials Balance for Benzene, Level 1 — Preliminary (EPA-560/13-80-014), prepared for the U.S. Environmental Protection Agency, McLean, Va., 1980; as cited in **IARC Monographs,** Vol. 29, 1982.
21. **Colenutt, B. A. and Thorburn, S.,** Gas chromatographic analysis of trace hydrocarbon pollutants in water samples, *Int. J. Environ. Stud.,* 15, 25, 1980.

Chapter 2

ANALYTICAL TECHNIQUES

Masayuki Ikeda and Toshihide Okuno

TABLE OF CONTENTS

I. INTRODUCTION*

The major route of exposure of humans to benzene is via inhalation. Accordingly, discussion in this chapter will be focused on the measurement of benzene in air.

Benzene in ambient air (community air or air in general environment) is known to be at ppb (10^{-9}, v/v) levels or even lower, and various other pollutants are also present at comparable levels. In contrast, benzene levels in the occupational environment is generally at ppm (10^{-6}, v/v) levels with a rather limited variety of other pollutants, if present. Currently, more attention has been paid to develop personal samplers (i.e., light and small sampling devices to be equipped by factory workers so that benzene in breathzone air will be sampled over a working time of the day) in addition to conventional stationary samplers.

The methods employed in these two major fields of application are generally different in response to the requirements. Thus, they will be described separately, i.e., Section II for the measurement of benzene in ambient air, and Section III for the measurement of benzene in the occupational environment.

II. MEASUREMENT OF BENZENE IN AMBIENT AIR**

A. General View

Literature on the methods of determining benzene in ambient air will be reviewed with a focus on gas chromatography (GC) and gas chromatography/mass spectrometry (GC/MS). Benzene in ambient air is mostly either from automobile exhaust or emitted in association with combustion of fuel oil. So-called background level of benzene in ambient air was summarized as in the range of 0.025 to 57 ppb.[1] For example, benzene was detected in community air in Europe at the level of 0.5 to 16.3 ppb.[2]

In the past 20 years, GC with flame ionization detectors (FID-GC) has been popularly used to measure benzene levels of 1 to several 10 ppb. Theoretically, benzene at such concentration should be detectable as the sensitivity of FID is in the range of 1 to 10 ng. However, in the case of community air analyses for benzene, careful selection of GC conditions is essential because varieties of other hydrocarbons are also present at comparable levels.[1] Mass spectrometry has been developed recently as a method, when combined with GC, of high sensitivity and selectivity. After being separated by GC, the mass number and molecular structure of each pollutant are determined by MS. There are two methods available in MS, namely mass chromatography (MC) and mass fragmentography (MF). In both methods, identification of benzene is achieved by means of computer-based data search system. The detection limit of benzene is 1 to 10 pg when m/z 78 fragment ion is utilized in MF.[3]

B. Measurement by Gas Chromatography

The relative response per mole of FID is approximately in proportion to the number of carbon atoms (carbon number) in the molecule of the hydrocarbon in concern.[4] For example, the relative response per mole of benzene (with a carbon number of 6) is 5.8 while that for *n*-butane (carbon number 4) is 4.0 (Table 1). The minimum detectable concentration by volume is 10^{-11} v/v for benzene.[5] When one 1-mℓ air sample is employed, the minimum detectable concentration of FID-GC was estimated to be 10 ppb for *n*-butane.[6]

Recently, Sexton and Westberg[7] observed in a study in the North American continent that benzene levels in the air are in the range of 4 to 18 ppb in seven urban centers, and <0.5 to 1.5 ppb in six rural areas. In their study, about 1000 air samples were analyzed by FID-GC. Air samples were collected in either polyfluoroethylene bags or stainless steel cans.

---

*   By M. Ikeda.
** By T. Okuno.

Table 1
RELATIVE RESPONSE OF DETECTOR

Compound	Relative response (per unit wt)	Relative response (per mol)	Carbon no.
Methane	3.4	0.95	1
Ethane	3.9	2.0	2
Ethylene	4.1	2.0	2
Acetylene	4.9	2.2	2
Propane	3.9	3.0	3
Propylene	4.0	2.9	3
Cyclopropane	4.3	3.1	3
Butane	4.0	4.0	4
Isobutane	4.0	4.0	4
1-Butene	4.1	4.0	4
2-Butene	4.0	3.9	4
Isobutylene	4.0	3.9	4
1,3-Butadiene	4.3	4.0	4
Hexane	4.2	6.3	6
Benzene	4.3	5.8	6
Cyclohexane	4.3	6.3	6
Heptane	4.3	7.4	7
Methanol	1.49	0.83	1
Ethyl alcohol	2.6	2.0	2
Carbon tetrachloride	0.24	0.64	1
Chloroform	0.40	0.85	1
Freon 12	0.19	0.40	1
Vinyl chloride	1.61	1.75	2
Carbon monoxide	0	0	1
Carbon dioxide	0	0	1
Nitrous oxide	0	0	0
Carbon disulfide	0	0	1

From Andreatch, A. J. and Feinland, R., *Anal. Chem.*, 32, 1021, 1960. With permission.

Lower molecular weight hydrocarbons (C_2 to C_5) were analyzed on a capillary column (1/16 mm × 20 m) packed with Dura-Pack® n-octane/Porasil® C, and higher molecular weight ones (C_6 to C_{10}) were on an SE-30 gas capillary column (1/16 mm × 30 m). However, the possibility that the hydrocarbons might either be adsorbed on the inner surface of the containers or decomposed there during transport and storage[8] should be taken into account.

In cases where the number of the samples to be collected are rather small, the active sampling method has often been used, in which pollutant-laden air is aspirated by a suitable pump and the pollutants are collected on an appropriate adsorbent such as charcoal[9] or porous polymer beads.[10,11] The trapped pollutants (including benzene) are subjected to GC analysis after thermodesorption or extraction with a suitable solvent (e.g., carbon disulfide for charcoal).

Taking advantage of diffusion, passive sampling devices with no pump to collect pollutants have recently been developed (for further details on passive sampling, see Chapter 2, III.E.). The devices were originally designed for occupational hygiene purposes, but the possible application to ambient air monitoring is currently under investigation. For example, Coutant and Scott[12] found that one commercially available passive sampler with a sampling rate of 71 cm³/min had a detection limit of 0.65 ppb.

In the study of Coutant and Scott,[12] detection of hydrocarbons (including benzene) was conducted with a photoionization detector (PID). As shown in Table 2, PID is over ten

Table 2

HYDROCARBON CLASS PID/FID NORMALIZED RESPONSES

	Retention times			
	<17 min		>17 min	
Class	Species tested	TNR[a] (mean ± SD)	Species tested	TNR[a] (mean ± SD)
Halogenated alkanes	9	0 ± 0	6	1 ± 1
Simple alkanes	13	3 ± 3	10	12 ± 4
Cycloalkanes and trimethylalkanes	2	3 ± 1	5	27 ± 8
Alkynes	3	3 ± 3	0	—
Alkenes	23	70[b] ± 11	20	55 ± 10
Aldehydes	4	69[c] ± 10	1	56
Ketones	2	157 ± 5	2	123 ± 25
Aromatics	0	—	21	87 ± 18
Chlorinated aromatics	0	—	7	141 ± 12
Chlorinated alkenes	3	218 ± 180	4	211 ± 97
Sulfur hydrocarbons	2	500 ± 210	2	129 ± 37

[a]   Toluene-normalized response (response to toluene = 100).
[b]   Does not include ethylene.
[c]   Does not include acetaldehyde.

From Cox, R. D. and Earp, R. F., *Anal. Chem.*, 54, 2265, 1982. With permission.

times more sensitive to aromatics than to simple alkanes.[13] When an automobile exhaust sample was analyzed by both PID-GC and FID-GC, benzene peak was clearly detected by PID, but not so by FID[13] (Figure 1). Sensitivity was also higher with PID than FID when an air sample from a parking lot was analyzed with the two detectors (Figure 2 and Table 3).[14]

Table 4 summarizes the relative response of PID and FID to various hydrocarbons when the geometries of the ion collector and the polarizing electrode were optimized.[14] Benzene had the highest PID/FID relative response among the chemicals studied, and the detection limit of benzene with PID was as low as 0.6 pg. By the combination of collection with the Porapak® N adsorption tube and PID-GC analysis, Van Tassel et al.[15] detected 7.2 ± 1.3 to 21.7 ± 1.5 pg benzene per cubic meter (mean ± SD, n = 3) in the air in the state of New York.

In comparing the results of air analyses from various institutions, possible interlaboratory error should be kept in mind. An interlaboratory precision study on benzene and other volatile organics, with the participation of four laboratories, disclosed that a bias of up to 46% from the standard took place in the accuracy of benzene determination (Table 5).[16] Another factor of error in evaluation is time-to-time change in benzene concentration in atmospheric air. Hester and Meyer[17] developed an automated PID-GC system in which the benzene concentration was measured at 10-min intervals. The application of the system to monitor benzene levels in the air of Newbury Park, Calif. disclosed a wide variation in the concentration; from 0.5 to 6.0 ppb during the 2-day study period (Figure 3).[17]

Recently, McClenny[18] established a system with an automated cryogenic preconcentration method. The system can automatically condense each of several 100-mℓ air samples in a small nickel tube by cooling with liquid nitrogen, and by heating the condensate is brought into an automated PID-GC. When 200 mℓ of air was analyzed, overall efficiency of collection and recovery for volatile organics was in the range of 100 ± 5%.

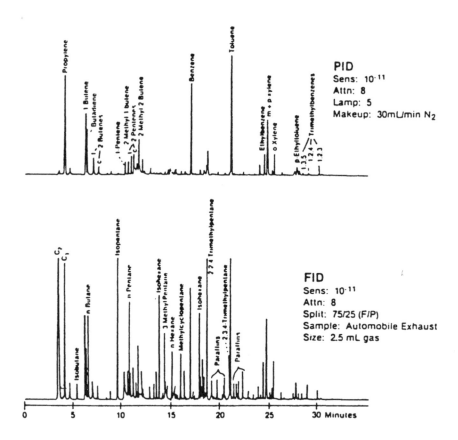

FIGURE 1. Comparison of both PID-(top) and FID-(bottom) GC of automobile exhaust (1981 American model). (From Cox, R. D. and Earp, R. F., *Anal. Chem.*, 54, 2265, 1982. With permission.)

C. Measurement with a Gas Chromatograph/Mass Spectrometer/Computer System (GC/MS/COMP)

The GC/MS/COMP system[19] has been introduced as a method to identify and determine trace amounts of benzene. Its accuracy and precision is better than that of conventional GC, because the system can identify benzene by means of molecular weight and structure, even in cases where benzene peak may overlap with those of other compounds on GC.

When the pollutants in the air were collected on Tenax® GC (7.96 mℓ, 35-60 mesh) in a Pyrex® glass sampling cartridge (1.6 cm in outer diameter and 10.5 cm in length), after Krost et al.,[20] and subjected to GC/MS/COMP analyses, it was found that benzene levels were 0.11 to 4.6 ppb in the ambient air, and 2.7 to 55 ppb in the air in the vicinity of industries.[3] The breakthrough air volume of the cartridge at 10, 21.1, and 32.2°C for benzene at ppb levels was reported to be 108, 54, and 27 ℓ, respectively.[20]

Repeated automated analyses of urban air for the entire month of April were conducted in Japan with FID-GC using Porapak® Q (6 g, 80-100 mesh) as the adsorbent.[21] Accordingly, it was found that the mean benzene concentration was 2.23 ppb with the minimum and the maximum of 0.1 and 8.1 ppb, respectively.[21] Oki et al.[22] developed an automated system to analyze volatile organics, in which the pollutants were adsorbed for 5 min on a Tenax® GC (0.2 g) tube at a flow rate of 200 mℓ/min and analyzed with a GC/quadropole-type MS (Shimadzu QP-1000)/COMP. The detection limit was 10 to 100 pg. Figure 4 illustrates typical mass fragmentograms of benzene and other aromatics in the air (1 ℓ collected) of the urban area, Kobe, Japan.[22]

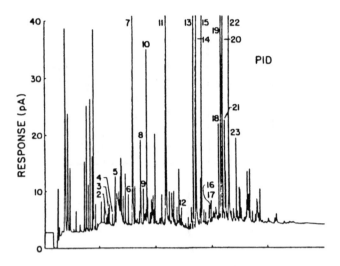

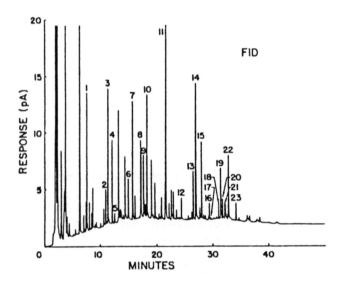

FIGURE 2. Comparison of both PID-(top) and FID-(bottom) GC of a parking lot air sample. An air sample, 500 mℓ, was collected in a small parking lot on February 4, 1983. The UV lamp intensity for PID was 70% full scale. The names of chemicals shown by numbers are given in Table 3. The peak for benzene is Peak No. 7. (From Nutmugal, W. and Cronn, D. R., *Anal. Chem.*, 55, 2160, 1983. With permission.)

Ross and Colton[23] reported on secondary ion mass spectrometry, utilizing the direct ionization of organics adsorbed on the surface of silver metal or graphitized carbon black. The method has a sensitivity to detect 2 ng benzene.

The mass fragment ion, m/z 78, has been employed for the analysis of benzene in ambient air by GC/MS/COMP system. A higher capacity to detect 1 pg benzene can be expected with a multichanneled double-focusing mass spectrometer, which can simultaneously monitor m/z 91 and 104 in addition, as shown in Figure 4.

Table 3
HYDROCARBON COMPOUNDS
DETERMINED IN THE REAL AIR
SAMPLE OF FIGURE 2 INCLUDING
THE OBSERVED PID/FID RATIOS
NORMALIZED TO TOLUENE

Peak no.	Compound	PID/FID normalized to toluene ($\times 100$)
1	*n*-Butane	
2	2,3-Dimethylbutane	4
3	2-Methylpentane	3
4	3-Methylpentane	3
5	1-Hexene	75
6	2,4-Dimethylpentane	15
7	Benzene	129
8	2,3-Dimethylpentane	40
9	3-Methylhexane	12
10	2,2,4-Trimethylpentane	28
11	Toluene	100
12	*n*-Octane	22
13	Ethylbenzene	84
14	*p*- and *m*-Xylene	116
15	*o*-Xylene	82
16	*n*-Nonane	25
17	Isopropylbenzene	134
18	*n*-Propylbenzene	103
19	*p*-Ethyltoluene	76
20	1,3,5-Trimethylbenzene	158
21	*o*-Ethyltoluene	105
22	1,2,4-Trimethylbenzene	123
23	1,2,3-Trimethylbenzene	95

From Nutmagul, W. and Cronn, D. R., *Anal. Chem.*, 55, 2160, 1983. With permission.

III. MEASUREMENT OF BENZENE IN OCCUPATIONAL ENVIRONMENT*

A. General View

In this section, the focus of the review will be on the methods of measuring atmospheric concentrations of benzene in occupational settings. The methods used for biological monitoring of exposure to benzene[24-28] will not be discussed. The readers are referred to a review on this particular topic, e.g., Lauwerys.[27]

Current methods employed to measure benzene in the air of workplaces (at ppm levels) can be classified as shown below.

1. Grab sampling of air followed by direct injection to a gas chromatograph (GC)
2. The detector tube method
3. The solid adsorbent method

Method 3 may be further divided into two subclassifications.

* By M. Ikeda.

Table 4
PID/FID RELATIVE RESPONSES AND NORMALIZED
PID/FID RESPONSES OF HYDROCARBON COMPOUNDS
FOR ELECTRODE GEOMETRY 1

Compound	PID/FID	PID/FID normalized to n-octane	PID/FID normalized to toluene ($\times 100$)	Min detection limit (pg) for the PID
n-Hexane	0.19 ± 0.03	0.42	2	14.7
n-Heptane	0.58 ± 0.03	1.29	7	9.0
n-Octane	0.45 ± 0.01	1.00	5	7.5
n-Nonane	0.72 ± 0.02	1.60	8	6.7
n-Decane	0.90 ± 0.02	2.00	10	5.1
1-Hexene	6.48 ± 0.12	14.4	74	1.6
2-Heptene	7.68 ± 0.16	17.1	88	1.3
1-Octene	4.73 ± 0.14	10.5	54	2.1
1-Nonene	4.30 ± 0.09	9.6	49	2.7
1-Decene	3.99 ± 0.21	8.9	46	2.1
Benzene	11.94 ± 0.27	26.5	137	0.6
Toluene	8.71 ± 0.22	19.4	100	0.8
Ethylbenzene	7.91 ± 0.17	17.6	91	1.3
o-Xylene	7.52 ± 0.10	16.7	86	1.3
Isopropylbenzene	6.03 ± 0.46	13.4	69	1.8

From Nutmagul, W. and Cronn, D. R., *Anal. Chem.*, 55, 2160, 1983. With permission.

3a. Active sampling methods which need pumps to aspirate air into adsorbent-containing tubes

3b. Passive (or diffusive) sampling methods which take advantage of diffusion onto the adsorbent

Table 6 classifies the methods from the viewpoints of sampling duration, application to personal/stationary sampling, necessity of analytical instrument, etc. Except for detector tubes, most methods use gas chromatography for the determination of benzene. Although colorimetry for benzene is also possible (e.g., Verma and Gupta),[29] the use of scrubbing liquids is inconvenient for personal sampling. In addition, color-forming reaction needs additional steps when interferants such as toluene are also present in air.

The most popular solid adsorbent is activated carbon (or so-called charcoal). The efficiency of benzene collection on activated carbon under experimental condition was reported to be 100%.[30] In both active and passive sampling, the benzene adsorbed on to activated carbon will be extracted with carbon disulfide;[31] high desorption efficiency of over 90% has been reported by various authors (e.g., 92%,[30] at least 93%,[32] 94%,[33] 96%,[34,35] and 98%[36]). When silica gel is used as the adsorbent, acetone is an extractant of choice with the efficiency of over 95% (Figure 5).[33] The selection of columns in GC analysis depends on the complexity of solvent mixture. U.S. National Institute of Occupational Safety and Health[37] recommends a 3-ft (ca. 1 m) × $^1/_8$-in. (ca. 3 mm) column packed with Porapak® Q (50-80 mesh) for FID-GC. Other columns (2 m × 3 mm; glass or stainless steel) recommended are those packed with either 5% polyethyleneglycol (PEG) 600, 10% PEG 20M, 10% Silicone DC 550, 10% squalane, or 10% tricresyl phosphate (TCP) on Chromosorb® WAW (60-80 mesh).[38] A CEF, i.e., n-Bis(2-cyanoethyl)formamide substrate on Chromosorb® P AW-DMCS (60-80 mesh) stainless steel column (1 m or 3.3 m × 3 mm)[39] and a 30% CEF on pink support column (2 m × 6 mm in outer diameter)[40] (Figure 6) were recommended for separation of low boiling point aromatics (such as benzene) from saturates and olefins in

Table 5
RESULTS OF VOST (ENTIRE TRAIN) VS. VOST CARTRIDGES

Activity audited	Audit gases	NBS[b] conc. ppb	Lab A Range (ppb)	Lab A Average[c] ppb	Lab A % acc.[d]	Lab B Range (ppb)	Lab B Average[c] ppb	Lab B % acc.[d]	Lab C[a] Range (ppb)	Lab C[a] Average[c] ppb	Lab C[a] % acc.[d]	Lab D Range (ppb)	Lab D Average[c] ppb	Lab D % acc.[d]
VOST (both sampling and analysis)	Carbon tetrachloride	21	5.9—15.3	10.4	−50	24.5—30.4	26.6	27	22.0—24.0	23.0	10	15—21	18.0	−14
	Chloroform	23	21.7—25.2	23.1	0	28.8—29.8	29.4	28	26.0—30.0	27.7	20	18—19	18.5	−20
	Perchloroethylene	29	29.7—36.3	33.9	17	35.4—41.4	38.8	34	Not audited			29—36	32.5	12
	Vinyl chloride	31	12.8—18.2	15.5	−50	7.8—13.0	10.5	−66	Not audited			24—32	28.0	−10
	Benzene	18	22.9—24.1	23.5	31	22.9—24.8	23.7	32	15.0—17.0	16.3	−9	19—19	19.0	6
VOST cartridges (analysis only)	Carbon tetrachloride	21	10.4—14.9	12.7	−40	22.6—24.0	23.3	11	22.0—27.0	23.7	13	19—20	19.5	−7
	Chloroform	23	23.6—25.0	24.3	6	25.8—27.0	26.4	15	26.0—38.0	30.3	32	18—19	18.5	−20
	Perchloroethylene	29	37.5—38.5	38.0	31	30.9—35.8	33.4	15	Not audited			33—36	34.5	19
	Vinyl chloride	31	13.1—20.8	16.4	−47	7.7—24.1	14.6	−53	Not audited			33—39	35.5	15
	Benzene	18	25.4—27.0	26.2	46	21.5—23.4	22.5	25	21.0—21.0	21.0	17	17—19	18.0	0

[a] Analyses by GC and not GC/MS.

[b] NBS values were obtained by direct GC analyses.

[c] Average based on duplicate or triplicate runs.

[d] $\% \text{ accuracy} = \dfrac{\text{laboratory value} - \text{NBS value}}{\text{NBS value}} \times 100$

From Jayanty, R. K. M., Sokash, J. A., Gutknecht, W. F., and Decker, C. E., *J. Air Pollut. Control Assoc.*, 35, 143, 1985. With permission.

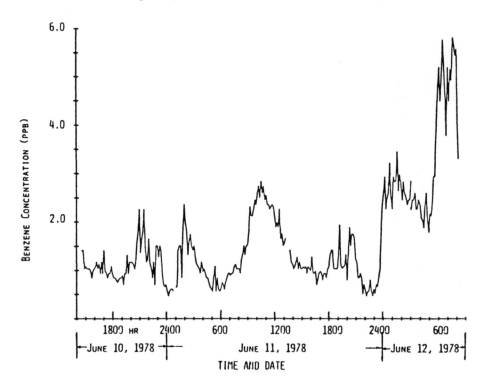

FIGURE 3.   Benzene concentrations in Newbury Park, Calif., as repeatedly measured for two days. (From Hester, N. E. and Meyer, R., *Environ. Sci. Technol.*, 13, 107, 1979. With permission.)

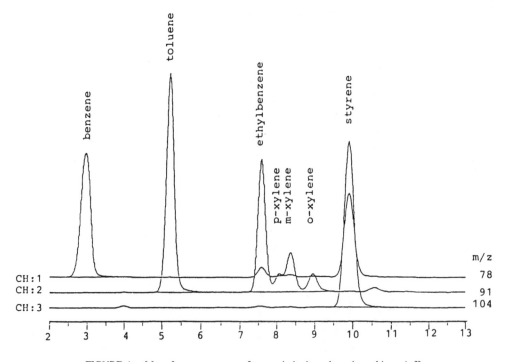

FIGURE 4.   Mass fragmentograms of aromatic hydrocarbons in ambient air.[22]

Table 6
METHODS APPLIED TO MEASURE BENZENE IN THE WORKPLACE AIR

Applicability	Grab sampling	Detector tubes	Sampling with solid adsorbents	
			Active	Passive
Short-term sampling (5 min)	Yes	Yes		
TWA measurement[a]		Yes[b]	Yes	Yes
Sampling devices (wt in g)	Hand pump (300 to 1000)	Hand pump (300 to 1000)	Electric pump (less than 1000)	Sampling case[c] (less than 50)
Analytical instrument	GC	None	Desorption and GC	Desorption and GC
Selectivity	Good	Poor[d]	Good	Good

[a]   4 to 8 hr weighted average; fit for both personal and stationary sampling.
[b]   Available at least from one commercial source.
[c]   No pump is needed; light weight is advantageous for sampling.
[d]   May cross-react with other solvent(s) when other solvent(s) coexists.

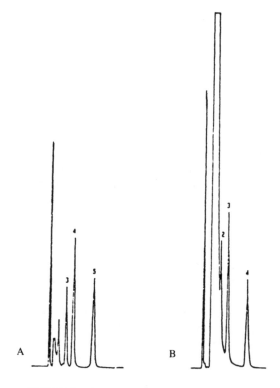

FIGURE 5.   Gas chromatograms of organic solvents eluted from charcoal with carbon disulfide or from silica gel with acetone; (A) Separation of charcoal tube eluates in CS$_2$, (1) n-Hexane (2) Methyl acetate (3) MEK (4) Benzene (5) Toluene; (B) Separation of silicagel tube eluates in acetone, (1) n-Hexane (2) MEK (3) Benzene (4) Toluene. (From Tada, O. and Cai, S.-X., J. Sci. Labour, 56, 453, 1980 [Japanese with an English abstract]. With permission.)

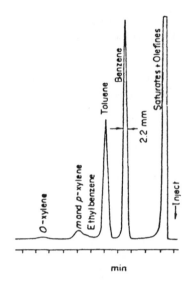

FIGURE 6.   A gas chromatogram of gasoline on a 30%
CEF on pink support column. Gasoline vapor, 0.5 mℓ,
was injected. Carrier gas, N$_2$ at 60 mℓ/min. Column
temperature at 100°C. (From Baxter, H. G., Blakemore,
R., Moore, J. P., Coker, D. T., and McCambley, W.
H., *Ann. Occup. Hyg.*, 23, 117, 1980. With permission.)

gasoline or petroleum distillates. The relative retention times of benzene and its alkyl derivatives on the latter GC column were described.[40] Furthermore, the relative retention times
of more than 100 organic solvents (including benzene) on nine glass columns have been
reported.[41] Capillary columns (e.g., 50 m × 0.25 mm Silicone OV 101 FS-WCOT)[42] may
be necessary for the analysis of benzene in a complex matrix (Figure 7). It should be noted
that high humidity (e.g., over 80%) may disturb adsorption of benzene on activated carbon,[36,43] although the performance may vary depending on the carbon preparation.[43]

It is recommended to refrigerate the carbon as soon after sample collection as possible.[36]
At 4°C, no significant change in benzene contents in carbon cloth packed in aluminum foil
was observed in 6-month periods.[44] In another experiment, less than 10% loss was observed
in toluene (methylbenzene) when the activated carbon cloth (packed in aluminum foil) was
kept at room temperature for 2 weeks, and the loss was more than 20% in *n*-hexane under
the same conditions. No loss in toluene nor *n*-hexane was detected in 2 weeks when kept
at 4°C.[45]

Regarding the validation of the method employed, Taylor et al.[46] recommended that the
overall accuracy of a sampling method in the range of 0.5 to 2.0 times the occupational
exposure standard should be ±25% for 95% of the samples tested. Based on the further
discussion of a possible statistical protocol (e.g., Shotwell et al.[47]), the overall accuracy can
be expressed as

$$\text{overall accuracy} = \pm\text{absolute mean bias} + (2 \times \text{mean CV})$$

if sufficient samples are tested.[35] Melcher et al.[36] proposed the criteria for successful validation as summarized in Table 7.

B. Direct Measurement

Benzene concentration at a given time can be measured by direct injection of the air grab-
sampled in either an airtight syringe or an air bag (from which a portion is taken into a

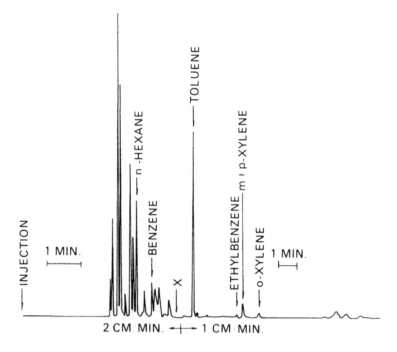

FIGURE 7. A gas chromatogram of automobile gasoline on a Silicone OV 101 FS-WCOT capillary column. (From Ikeda, M., Kumai, M., Watanabe, T., and Fujita, H., *Ind. Health*, 22, 235, 1984. With permission.)

Table 7
CRITERIA FOR SUCCESSFUL VALIDATION

1. Accuracy and precision of combined analytical and collection procedures for concentration range of 0.1 × TLV to 2 × TLV — ±16% relative at the 95% confidence level
2. Total recovery efficiency of at least 75%
3. Bias between total recovery and desorption efficiency less than ±10% relative
4. Capable of taking 10- to 15-min samples at ceiling and excursion levels
5. Minimum sampling time 1 hr, preferred 4 to 8 hr for TWA
6. Storage samples should compare within ±10% relative to initial samples after being stored for 14 days
7. The flow rate of the sampling pump should be known with the accuracy of ±5%

From Melcher, R. G., Langer, R. R., and Kagel, R. O., *Am. Ind. Hyg. Assoc. J.*, 39, 349, 1978. With permission.

syringe and injected into GC). In the latter case, care should be taken to avoid possible adsorption on and penetration through the bag. For this reason, the use of bags made of aluminum foil-plastics laminate[48] or polyfluoroethylene (Tedlar)[49] are recommended.

In addition to conventional FID-GC (which uses flame, needs hydrogen and nitrogen gas containers, as well as air supply, and is heavy in weight), a portable GC with a photoionization detector (PID-GC) has become commercially available.[50,51] The instrument is explosion-safe, as it does not use flame for detection, does not need hydrogen supply, weighs about 10 kg, and can detect benzene at the levels from ppb to ppm range by direct injection of air samples.[50] The PID is most sensitive to aromatics, followed by olefin hydrocarbons and

Table 8
EXAMPLES OF DETECTOR TUBES FOR BENZENE
FROM VARIOUS COMMERCIAL SOURCES

Brand	Model	Measurable range (ppm)	Sampling volume of air (number of strokes[a])
Dräger®	0.5/a(67 28651)	0.5—10	2—40
	5/a(67 18801)	5—40	2—15
	5/b(67 28071)	5—50	20
	0.05(CH 24801)	0.05—1.4 mg/ℓ	2—20
Gastec®	No. 121	5—120	1—4
	No. 121L	0.25—60	1—10
Kitagawa®	118SB	5—200	1
	118SC	1—100	1—4
MSA®	460754	5—100	1—6
	93074	5—200	1—3

[a]   100 mℓ per stroke.

then paraffin hydrocarbons,[49] but the sensitivity may be influenced by high humidity (e.g., 90%[52]). Experiences with this instrument[53] indicate that introduction of a heating/thermostabilizing system will improve the performance by eliminating the drift of chromatogram baseline due to changes in the room temperature. An extension switch to reduce sensitivity should be installed if a 1-mℓ syringe is used, while a microsyringe would be readily spoilt, e.g., by paint droplets in air sampled in a spray-painting room.

C. The Detection Tube Method

The detector tubes have a history of more than 20 years since Kitagawa[54] made the presentation on this method. The tubes have been providing a rapid, inexpensive, and simple method for evaluating the levels of various gaseous pollutants (including benzene) in the industrial environment. Ash and Lynch[55] examined the reliability of commercially available benzene detector tubes manufactured for the vapor concentration range of 20 to 160 ppm and found that, while no tubes met the criteria of "within ± 25% of the stated concentration over the working range of the tube", tubes from three manufacturers were found to be ± 50% accurate at the 95% confidence limit when examined in air at 50% relative humidity. Currently, tubes are available from several commercial sources under various brands including Dräger®, Gastec®, Kitagawa®, and MSA® (Table 8). While the original detector tubes were designed to measure benzene concentrations in the air grab-sampled with hand pumps in a short term of 3 to 5 min (either in the breathzone of workers or at a fixed station), long-term detector tubes (including that for benzene) using portable pumps for aspiration of the air have been developed for personal sampling to cover a 1- to 4-hr period. A laboratory test revealed that, although certain improvement remains to be made to increase reliability, the precision and accuracy of the long-term tubes may encourage the use for monitoring personal exposure.[56]

The detector tubes are mainly used under normal climatical conditions. Tests with tubes for CO and H_2S under extreme conditions revealed that high (e.g., 650°C) and low (e.g., − 20°C) temperatures cause systematic plus and minus indications, respectively, and the reading is directly proportional to the absolute air pressure, while high or low humidity does not influence the results.[57] Similarly, the reading of benzene long-term tubes was lower at over 40% relative humidity than at 30% relative humidity.[56]

One of the major problems associated with the use of detector tubes is the possible cross-reaction of the color-generating reagent in the tube with other coexisting solvents,[58] which jeopardizes the reading. It was reported that benzene tubes of one brand responded to toluene with a sensitivity of about 10%.[56]

D. The Charcoal Tube Method

The U.S. National Institute for Occupational Safety and Health[37,59] published manuals in which a method was described for the determination of benzene in the occupational environment. In principle, a known volume of air is drawn through a glass tube (7 cm in length and 4 mm in inner diameter) containing 20-40 mesh activated charcoal in two sections (100 mg in the first section and 50 mg in the backup section, separated by a 2-mm portion of urethane foam [so-called NIOSH charcoal tube]) by means of a either a personal pump (for personal sampling) or any vacuum pump (for stationary sampling). The recommended volume of air for sampling is 2 ℓ for determination of a 10-min ceiling, and 12 ℓ for the measurement of an 8-hr time weighted average concentration. Commercial personal pumps are available to sample at flow rates as low as 1.0 mℓ/min.[9] After sampling of air is terminated, the benzene in the carbon of the first section is desorbed into 1.0 mℓ carbon disulfide, an aliquot of which is subjected to gas chromatography. The carbon of the backup section is similarly analyzed. The sum of the two after correction for the corresponding blank is taken as the total measured amount. The detection of a significant amount of benzene in the backup section indicates that the first section is overloaded. When the amount of benzene in the backup section exceeds 25% of that found in the first section, the possibility of sample loss exists.[60] The breakthrough capacity of the NIOSH-type tube for benzene was estimated to be 6 mg/100 mg charcoal, and the suggested maximum sample volume was 10 ℓ.[60] Immediately after sampling, both ends of each charcoal tube should be capped tightly; no significant loss of benzene in the tube was observed over 2 weeks, even when kept at room temperature, using either polyethylene or polyvinyl chloride caps,[32] while the use of rubber caps must be avoided.[37] When the NIOSH tubes were exposed to benzene at 8 levels (3.1 to 24.0 ppm) in air at 25°C and 80% relative humidity for 6 hr at 50 cm^3/min, the average relative error was +0.8%; the relative error ranged from −24.8% to +20.2% with a smallest absolute value of 2.8% at 7.5 ppm (Table 9).[61] A performance test of the NIOSH tubes with several organic solvents, including toluene (methylbenzene), revealed that humidity has a pronounced effect on the breakthrough capacity of toluene, that toluene is stable on the tube for more than 2 weeks when kept at room temperature, and does not migrate from the first section to the backup section during the storage, and that the desorption efficiency is dependent on the charcoal lot.[9] In a collaborative testing study, the overall error (i.e., the sum of sampling and analytical errors) for benzene determination was 13.35% (relative error) with roughly equal shares for sampling error (major source being the error in pump calibration) and for analytical error (with significant portion in determination of desorption factors).[62]

It is also possible to use adsorbents other than charcoal, e.g., silica gel in combination with acetone as desorbing liquid.[33] A combination of a vapor adsorption tube (packed with either Porapak® Q, Tenax® GC or coconut charcoal) — heat desorption — FID-GC was presented as a method of high sensitivity with no need for sample preparation.[40] A refinement of the NIOSH method was also described.[40]

E. The Passive Sampling (or Diffusive Sampling) Method

So-called passive samplers take advantage of a diffusion mechanism (i.e., without any pump) for sampling of volatile chemicals such as organic solvents including benzene. The theory for the diffusion mechanism has also been well developed.[63,64] Samplers of various types have been developed for the determination of time-weighted average concentration (e.g., for 8 hr) of organic solvents including benzene. They are applicable both as personal samplers and as stationary samplers. Currently, several types of samplers[65] are commercially

Table 9
PERFORMANCE OF NIOSH CHARCOAL TUBES FOR SAMPLING BENZENE

Benzene vapor (ppm)	Analytical results		
	Calculated (mg)	Found (mg)	Error (%)
3.1	0.162	0.156	− 3.7
7.5	0.352	0.362	+ 2.8
15.2	0.795	0.872	+ 9.7
19.6	0.990	1.190	+ 20.2
23.1	1.450	1.090	− 24.8
24.0	1.080	1.150	+ 6.5
24.0	1.300	1.090	+ 16.2
Mean			+ 0.8

Note: Sampling was conducted for 6 hr at 50 cm³ of air per minute at 25°C, 80% relative humidity.

From Kring, E. V., Ansul, G. R., Henry, T. J., Morello, J. A., Dixon, S. W., Vasta, J. F., and Hemingway, R. E., *Am. Ind. Hyg. Assoc. J.*, 45, 250, 1984. With permission.

available with various brands including Abcor Gasbadge (now National Mine Service),[66] 3M® 3500 Organic Vapor Monitor,[67] Du Pont® Pro-Tek Organic Vapor G-AA Air Monitoring Badge,[35] Dräger® ORSA 5,[68] and Porton Down Charcoal Cloth Diffusive Sampler.[69] While some of them employ activated carbon cloth as an adsorbent and therefore are in the shape of a badge,[35,66,67,69] others use carbon grains packed in a cylindrical holder (both ends of the tube to be used as adsorbing surfaces)[68] or graphitized carbon black (Carbopack B, 20-40 mesh) in a Pyrex® glass tube.[70] Other proposed materials include natural rubber disks (3.75 mm × 12 mm diameter),[71] Tenax® GC,[72] and Chromosorb® porous polymers.[73] In case activated carbon or natural rubber was used as an adsorbent, benzene adsorbed was extracted with carbon disulfide, and an aliquot of the carbon disulfide was subjected to GC analysis.[35,66-69,71] Thermal desorption was employed in the cases of Tenax® GC[72] and Carbopack B.[70]

The data published so far indicate that performance of passive samplers meets U.S. regulatory requirements for the accuracy of devices for monitoring exposure of workers,[66,67,74] that close correlation can be established between the results of those by the charcoal tubes and by the passive samplers,[74,75] and that no consistent difference was demonstrated in means between the charcoal tube results and passive sampler results in a field study in which benzene concentration was mostly less than 1 ppm.[76] A further variation of diffusion-type dosimeters is a stain length passive dosimeter (Figure 8).[77] When iodine pentoxide in 96% sulfuric acid was used as the color-generating reagent in Chromosorb® W and 6 mm-inside-diameter glass tubes (sealed at one end) were employed to hold the reagent, a curvilinear relationship was established between exposure (i.e., 4 to 21 ppm benzene for several hours) and stain length (cm).[77]

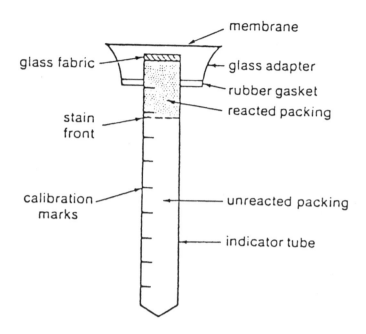

FIGURE 8. Schematic drawing of a stain length passive dosimeter. (From Sefton, M. V., Kostas, A. V., and Lombardi, C., *Am. Ind. Hyg. Assoc. J.*, 43, 820, 1982. With permission.)

REFERENCES

1. **Graedel, T. E.,** *Chemical Compounds in the Atmosphere,* Academic Press, New York, 1978.
2. **Coulston, F.,** Benzene-interpretation of data and evaluation of current knowledge, minutes of the workshop, *Regul. Toxicol. Pharmacol.,* 1, 244, 1981.
3. **Pellizzari, E. D.,** Analysis for organic vapor emissions near industrial and chemical waste disposal sites, *Environ. Sci. Technol.,* 16, 781, 1982.
4. **Andreatch, A. J. and Feinland, R.,** Continuous trace hydrocarbon analysis by flame ionization, *Anal. Chem.,* 32, 1021, 1960.
5. **Lovelock, J. E.,** Ionization methods for the analysis of gases and vapors, *Anal. Chem.,* 33, 162, 1961.
6. **Feinland, R., Andreatch, A. J., and Coutrupe, D. P.,** Automotive exhaust gas analysis by gas-liquid chromatography using flame ionization detection, *Anal. Chem.,* 33, 991, 1981.
7. **Sexton, K. and Westberg, H.,** Nonmethane hydrocarbon composition of urban and rural atmospheres, *Atmos. Environ.,* 18, 1125, 1984.
8. **Makide, Y., Kanai, Y., Tominaga, T.,** Background atmospheric concentrations of halogenated hydrocarbons in Japan, *Bull. Chem. Soc. Jpn.,* 53, 2681, 1980.
9. **Saalwaechter, A. T., McCammon, C. S., Jr., Roper, C. P., and Carlberg, K. S.,** Performance testing of the NIOSH charcoal tube technique for the determination of air concentration of organic vapors, *Am. Ind. Hyg. Assoc. J.,* 38, 476, 1977.
10. **Williams, F. and Umstead, M. E.,** Determination of trace contaminations in air by concentrating on porous polymer beads, *Anal. Chem.,* 40, 2232, 1968.
11. **Adams, J., Menzies, K., and Levins, P.,** Selection and Evaluation of Sorbent Resins for the Collection of Organic Compounds, U.S. Environmental Protection Agency, EPA-600/7-77-044, Washington, D.C., 1977.
12. **Coutant, R. W. and Scott, D. R.,** Applicability of passive dosimeter for ambient air monitoring of toxic organic compounds, *Environ. Sci. Technol.,* 16, 410, 1982.
13. **Cox, R. D. and Earp, R. F.,** Determination of trace level organics in ambient air by high-resolution gas chromatography with simultaneous photoionization and flame ionization, *Anal. Chem.,* 54, 2265, 1982.
14. **Nutmagul, W. and Cronn, D. R.,** Photoionization/flame-ionization detection of atmospheric hydrocarbons after capillary gas chromatography, *Anal. Chem.,* 55, 2160, 1983.

15. **Van Tassel, S., Amalfitano, N., and Narang, R. S.,** Determination of arenes and volatile haloorganic compounds in air at microgram per cubic meter levels by gas chromatography, *Anal. Chem.*, 53, 2130, 1981.

16. **Jayanty, R. K. M., Sokash, J. A., Gutknecht, W. F., and Decker, C. E.,** Quality assurance for principal organic hazardous constituents (POHC) measurements during hazardous waste trial burn tests, *J. Air Pollut. Control Assoc.*, 35, 143, 1985.

17. **Hester, N. E. and Meyer, R.,** A sensitive technique for measurement of benzene and alkylbenzenes in air, *Environ. Sci. Technol.*, 13, 107, 1979.

18. **McClenny, W. A.,** Automated cryogenic preconcentration and gas chromatographic determination of volatile organic compounds in air, *Anal. Chem.*, 56, 2947, 1984.

19. **Bunch, J. E., Castillo, N. P., Smith, D., Bursey, D., and Pellizzari, E. R.,** Evaluation of the Basic GC/MS Computer Analysis Technique for Pollutant Analysis, U.S. Environmental Protection Agency, Contract No. 68-02-2998, Washington, D.C., 1980.

20. **Krost, K. J., Pellizzari, E. D., Walburn, S. G., and Hubbard, S. A.,** Collection and analysis of hazardous organic emissions, *Anal. Chem.*, 54, 810, 1982.

21. **Hanai, Y., Katou, T., and Aoki, S.,** Automatic determination of C_6-C_9 aromatic hydrocarbons in the air, *Yokohama Natl. Univ. Environ. Res. Cent. Annu. Rep.*, 10, 23, 1983 (Japanese).

22. **Oki, N., Yoshioka, M., and Okuno, T.,** Automatic analysis of organic compounds in the atmosphere by GC/MS, presented at 26th Annu. Meet. of Japan Society of Air Pollution, Abstract No. 239, Tokyo, November 12 to 14, 1985 (Japanese).

23. **Ross, M. M. and Colton, R. J.,** Carbon as a sample substrate in secondary ion mass spectrometry, *Anal. Chem.*, 55, 150, 1983.

24. **Docter, J. H. and Zielhuis, R.,** Phenol excretion as a measure of occupational exposure, *Ann. Occup. Hyg.*, 10, 317, 1967.

25. **Angerer, J., Szadkowski, D., Manz, K., Patt, R., and Lehnert, G.,** Chronische Lösungsmittelbelastung an Arbeitsplatz, *Int. Arch. Arbeitsmed.*, 31, 1, 1973.

26. **Åstrand, I.,** Uptake of solvents in the blood and tissue of man: a review, *Scand. J. Work Environ. Health*, 1, 199, 1975.

27. **Lauwerys, R.,** Human biological monitoring of industrial chemicals. I. Benzene, Commission of the European Communities, Luxembourg, 1979, 1.

28. **Braier, L., Levy, A., Dror, K., and Pardo, A.,** Benzene in blood and phenol in urine in monitoring benzene exposure in industry, *Am. J. Ind. Med.*, 2, 119, 1981.

29. **Verma, P. and Gupta, V. K.,** A new method for spectrophotometric determination of benzene in air, *Environ. Int.*, 10, 279, 1984.

30. **Sidhu, K. S.,** An assessment of collection efficiency of some environmental contaminants on activated carbon, *Arch. Environ. Contam. Toxicol.*, 12, 747, 1983.

31. **Pannwitz, K.-H.,** Diffusion coefficients, *Dräger Rev.*, 52, 1, 1984.

32. **Fukabori, S., Tada, O., and Sugai, T.,** Determination of organic solvent vapours in air, *J. Sci. Labour*, 57, 11, 1981 (Japanese).

33. **Tada, O. and Cai, S.-X.,** Determination of organic solvents in air using charcoal tube and silicagel tube, *J. Sci. Labour*, 56, 453, 1980 (Japanese with an English abstract).

34. **Klingner, T. D.,** Charcoal tube user guide, 2nd ed., MDA Scientific, Inc., Park Ridge, Ill., 1979; as cited in **Sidhu, K. S.,** *Arch. Environ. Contam. Toxicol.*, 12, 747, 1983.

35. **Lautenberger, W. J., Kring, E. V., and Morello, J. A.,** A new personal badge monitor for organic vapors, *Am. Ind. Hyg. Assoc. J.*, 41, 737, 1980.

36. **Melcher, R. G., Langer, R. R., and Kagel, R. O.,** Criteria for the evaluation of methods for collection of organic pollutants in air using solid sorbents, *Am. Ind. Hyg. Assoc. J.*, 39, 349, 1978.

37. NIOSH Manual of Analytical Methods, Publ. No. (NIOSH) 75-121, National Institute for Occupational Safety and Health, U.S. Department of Health, Education and Welfare, Atlanta, 1974, 127-1.

38. Ministry of Labor, Japan, *Guidebook for Work Environment Measurements*, Vol. 2, 1980, 284 (Japanese).

39. **Esposito, G. G. and Jacobs, B. W.,** Chromatographic determination of aromatic hydrocarbon in ambient air, *Am. Ind. Hyg. Assoc. J.*, 38, 401, 1977.

40. **Baxter, H. G., Blakemore, R., Moore, J. P., Coker, D. T., and McCambley, W. H.,** The measurement of air borne benzene vapour, *Ann. Occup. Hyg.*, 23, 117, 1980.

41. **Sakai, K., Mitani, K., Tsuchiya, H., and Nakata, T.,** Studies on the environmental analyses in the workshops. IV. Identification of organic solvent by gas chromatography, *Nagoya City Inst. Health Annu. Rep.*, 29, 60, 1983 (Japanese).

42. **Ikeda, M., Kumai, M., Watanabe, T., and Fujita, H.,** Aromatic and other contents in automobile gasoline in Japan, *Ind. Health*, 22, 235, 1984.

43. **Hirayama, T. and Ikeda, M.,** Applicability of activated carbon felt to the dosimetry of solvent vapor mixture, *Am. Ind. Hyg. Assoc. J.*, 40, 1091, 1979.

44. **Inoue, O., Kasahara, M., Nakatsuka, H., Watanabe, T., Yin, S.-N., Li, G.-L., Jin, C., and Ikeda, M.,** Quantitative relation of urinary phenol level to breathzone benzene concentrations: a factory survey, *Br. J. Ind. Med.*, 43, 692, 1986.

45. **Ikeda, M., Kumai, M. and Aksoy, M.,** Applicability of carbon felt dosimetry to field studies distant from analytical laboratory, *Ind. Health*, 22, 53, 1984.

46. **Taylor, D. G., Kupel, R. E., and Bryant, J. M.,** Documentation of the NIOSH Validation Tests, Publ. No. (NIOSH) 77-185, U.S. Department of Health, Education and Welfare, Atlanta, 1977.

47. **Shotwell, H. P., Caporossi, J. C., McCollum, R. W., and Mellor, J. F.,** A validation procedure for air sampling — analysis systems, *Am. Ind. Hyg. Assoc. J.*, 40, 737, 1979.

48. **Gage, J. C., Lagesson, V., and Tunek, A.,** A method for the determination of low concentrations of organic vapours in air and exhaled breath, *Ann. Occup. Hyg.*, 20, 127, 1977.

49. **Tanaka, T. and Shinozaki, M.,** Application of PID gas chromatography to analysis of organic pollutants in the atmosphere, *Kogai*, 18, 31, 1983 (Japanese with an English abstract).

50. **Barker, N. J. and Leveson, R. C.,** A portable photoionization GC for direct air analysis, *Am. Lab.*, 12(12), 76, 1980.

51. **Leveson, R. C. and Barker, N. J.,** A portable multi-component air impurity analyzer having sub-part per billion capability without sample pre-condensation, *Anal. Instrum.*, 19, 7, 1981.

52. **Barsky, J. B., Que Hee, S. S., and Clark, C. S.,** An evaluation of response of some portable direct-reading 10.2 eV and 11.8 eV photoionization detectors and a flame ionization gas chromatograph for organic vapors in high humidity atmospheres, *Am. Ind. Hyg. Assoc. J.*, 46, 9, 1985.

53. **Ikeda, M., Watanabe, T., Kasahara, M., Kamiyama, S., Suzuki, H., Tsunoda, H., and Nakaya, S.,** Organic solvent exposure in small scale industries in northeast Japan, *Ind. Health*, 23, 181, 1985.

54. **Kitagawa, T.,** The rapid measurement of toxic gases and vapour, in Proc. 13th Int. Congr. Occup. Health, New York, 1960, 506.

55. **Ash, R. M. and Lynch, J. R.,** The evaluation of gas detector tube system: benzene, *Am. Ind. Hyg. Assoc. J.*, 32, 410, 1971.

56. **Jentzsch, D. and Fraser, D. A.,** A laboratory evaluation of long term detector tubes: benzene, toluene, trichloroethylene, *Am. Ind. Hyg. Assoc. J.*, 42, 810, 1981.

57. **Leichnitz, K.,** Use of detector tubes under extreme conditions (humidity, pressure, temperature), *Am. Ind. Hyg. Assoc. J.*, 38, 707, 1977.

58. **Tokunaga, R., Takahata, S., Onoda, M., Ishi-i, T., Sato, K., Hayashi, M., and Ikeda, M.,** Evaluation of the exposure to Organic solvent mixture, *Int. Arch. Arbeitsmed.*, 33, 257, 1974.

59. NIOSH Manual of Analytical Methods, 2nd ed., Vol. 3, HEW Publ. No. (NIOSH) 77.157.C, National Institute for Occupational Safety and Health, U.S. Department of Health, Education and Welfare, 1977, s311-1.

60. **Sawicki, E., Belsky, T., Friedel, R. A., Hyde, D. L., Monkman, J. L., Rasmussen, R. A., Ripperton, L. A., and White, L. D.,** Organic solvent vapors in air analytical method, *Health Lab. Sci.*, 12, 394, 1975.

61. **Kring, E. V., Ansul, G. R., Henry, T. J., Morello, J. A., Dixon, S. W., Vasta, J. F., and Hemingway, R. E.,** Evaluation of the standard NIOSH type charcoal tube sampling method for organic vapors in air, *Am. Ind. Hyg. Assoc. J.*, 45, 250, 1984.

62. **Larkin, R. L., Crable, J. V., Catlett, L. R., and Seymour, M. J.,** Collaborative testing of a gas chromatographic charcoal tube method for seven organic solvents, *Am. Ind. Hyg. Assoc. J.*, 38, 543, 1977.

63. **Palmes, E. D. and Gunnison, A. F.,** Personal monitoring device for gaseous contaminants, *Am. Ind. Hyg. Assoc. J.*, 32, 78, 1971.

64. **Posner, J. C. and Moore, G.,** A thermodynamic treatment of passive monitors, *Am. Ind. Hyg. Assoc. J.*, 46, 277, 1985.

65. **Feigley, C. E. and Chastain, J. B.,** An experimental comparison of three diffusion samples exposed to concentration profiles of organic vapors, *Am. Ind. Hyg. Assoc. J.*, 43, 227, 1982.

66. **Gillespie, J. C. and Daniel, L. B.,** A new sampling tool for monitoring exposures to toxic gases and vapors, Proc. Symp. Dev. Usage Pers. Monit. Exposure Health Eff. Stud., Publ. No. EPA-600/9-79-032, U.S. Environmental Protection Agency, Washington, D.C., 1979, 479.

67. **Gosselink, D. W., Braun, D. L., Mullins, H. E., and Rodriguez, S. T.,** A new personal organic vapor monitor with in situ sample elution, Proc. Symp. Dev. Usage Pers. Monit. Exposure Health Eff. Stud., Publ. No. EPA-600/9-79-032, U.S. Environmental Protection Agency, Washington, D.C., 1979, 365.

68. **Pannwitz, K.-H.,** A new sampling device for vapours of organic solvents, *Dräger Rev.*, 48, 8, 1981.

69. **Purnell, C. J., Wright, M. D., and Brown, R. H.,** Performance of the Porton Down charcoal cloth diffusive sampler, *Analyst (London)*, 106, 590, 1981.

70. **Bertoni, G., Perrino, C., Fratarcangeli, R., and Liberti, A.,** Critical parameters for the adsorption of gaseous pollutants on passive samplers made of low specific area adsorbents, *Anal. Lett.*, 18(A4), 429, 1985.

71. **Sefton, M. V., Kostas, A. V., and Lombardi, C.,** Stain length passive dosimeters, *Am. Ind. Hyg. Assoc. J.,* 43, 820, 1982.
72. **Coutant, R. W.,** Passive sampling devices with reversible adsorption, *Anal. Chem.,* 57, 219, 1985.
73. **Collenutt, B. A. and Jamal, A. R. A.,** Sorption studies of chromosorb porous polymers and their potential use in passive monitors, *Analyst,* 110, 273, 1985.
74. **Kring, E. V., Graybill, M. W., Morello, J. A., Ansul, G. R., Adkins, J. E., Jr., and Lautenberger, W. J.,** PRO-TEC organic vapor air monitoring badges, *ASTM Data Ser.,* 786, 85, 1982.
75. **Pannwitz, K.-H.,** Sampling and analysis of organic solvent vapours in the atmosphere: comparison of activity and passive sampling devices, *Dräger Rev.,* 52, 19, 1984.
76. **Hickey, J. L. S. and Bishop, C. C.,** Field comparison of charcoal tubes and passive vapor monitor with mixed organic vapors, *Am. Ind. Hyg. Assoc. J.,* 42, 264, 1981.
77. **Sefton, M. V., Mastracci, E. L., and Mann, J. L.,** Rubber disk passive monitor for benzene dosimeter, *Anal. Chem.,* 53, 458, 1981.

Chapter 3

EXPERIMENTAL BENZENE TOXICITY

Carroll A. Snyder

TABLE OF CONTENTS

I. INTRODUCTION

Any discussion of experimental benzene toxicity cannot be all encompassing. The literature is just too vast. In 1977, for example, a monograph was published listing over 1000 references concerning the toxicity of benzene.[1] Since then, there has been a dramatic increase in the publication rate of studies dealing with benzene toxicity. Rather than give a litany of publications, this review will emphasize those studies which illuminate the nature and the mechanism of benzene-induced toxic responses. The foundation of such a treatment consists of the observation, repeated often during 70 years of experimental benzene research, that most of the toxic responses produced by benzene in humans can be reproduced in animals. These include responses to short-term, intense exposures such as petechial hemorrhaging and central nervous system (CNS) toxicity,[2,3] as well as responses to much lower concentrations which are usually limited to hematopoietic toxicity. Because benzene is regulated at concentrations far below those required to induce CNS toxicity and hemorrhaging, these effects are infrequently studied now. It is in the area of hematotoxicity where most studies of benzene are concentrated and it is this area that will form the bulk of this review.

Perhaps the major impetus for the study of benzene toxicity has been the association between benzene exposure and the development of hematopoietic tumors in humans. Although benzene toxicity has been studied for over 70 years, successful experiments demonstrating its carcinogenicity have occurred only recently. In view of the importance of these developments, the area of experimental carcinogenesis will also be reviewed extensively.

II. HEMATOTOXICITY

The widespread use of benzene as a solvent and as a starting material for chemical synthesis began in the middle of the 19th century.[4] By the start of the 20th century, the first clinical reports of the hematotoxic effects of benzene appeared.[5-7] The most prevalent hematopoietic disorder found in benzene-intoxicated individuals was pancytopenia or one of its variants (depressions in the numbers of all or some of the circulating blood cell types). In some cases hematopoietic damage was so severe as to be fatal.[8] Almost concurrent with these developments were the first experimental studies of benzene toxicity.[9,10] Rabbits administered benzene parenterally exhibited leukopenia accompanied by marrow aplasia. During the course of repeated benzene treatments, white cell counts declined in sinusoidal patterns of decreasing amplitude. When treatments were discontinued, white cell counts rose in sinusoidal patterns of increasing amplitude.[10] Thus, shortly after the discovery of the hematotoxic effects of benzene in humans, it was determined that at least some of the cytopenic effects of benzene could be reproduced and studied in animals.

Many experimental studies of benzene toxicity have been concerned with dose/response and time-course/response relationships between benzene exposures and their effects on circulating blood cells.[11-18] Although these studies have employed a variety of animal species, dosages of benzene, and routes of exposure, certain consistent patterns have emerged. Animals exposed to toxic concentrations of benzene have usually exhibited lymphocytopenia as the initial response.[14-18] In some cases, it has been the only cytopenic response observed.[14-16] Depressions in circulating red cell levels have also been a common finding in experimental benzene exposures.[16-19] Anemia, however, has usually been observed following lymphocytopenia and has usually required higher benzene concentrations or longer treatments.[16-18] In direct contrast, granulocyte levels have been observed to markedly increase in response to benzene exposures.[16-18] Although granulocytopenia has been observed after short-term, intense benzene exposures,[19,20] protracted exposure regimes have demonstrated that granulocytes not only recover from the initial benzene insult, but markedly increase in number while lymphocyte and red cell numbers remain depressed.[16-18]

Thus, the major circulating blood cell populations respond differently to experimental benzene exposures. Lymphocytes and red cells decline with lymphocytes showing greater sensitivity to a given benzene treatment. Granulocytes, however, not only resist benzene treatment, but markedly increase during protracted exposures.

A. Lymphocytes

The particular sensitivity of lymphocytes to benzene exposure makes them attractive models for the study of benzene toxicity at the cellular level. In addition, the functional and/or proliferative capacities of large cohorts of lymphocytes can be measured by several in vitro assays. These assays are sensitive measures of cellular integrity since they require populations of lymphocytes to recognize and respond to various stimuli. With these assays, even subtle changes in cellular activity induced by benzene can be reliably measured.

Much of the information regarding the effects of benzene on lymphocyte populations has been provided by Irons and colleagues.[21-25] They have shown that the parenteral administration of benzene or some of its metabolites reduces the responses of lymphocyte populations to the proliferative stimuli of mitogens and to the antigenic stimulus of sheep red blood cells.[22,25] They have provided dose/response relationships for these effects and have shown that the benzene-induced reduction in mitogenic response can be ameliorated by a modulator of benzene metabolism.[22] They have also provided evidence that some metabolites of benzene act directly on certain lymphocyte populations in a manner similar to compounds known to disrupt cytoskeleton function.[23,24] From this latter observation, they have proposed a mechanism whereby metabolites of benzene capable of forming quinone structures disrupt cytoskeleton function by reacting with free sulfhydryl groups on proteins required for microtubule assembly.[23,26] In this way, benzene metabolites could prevent cells from completing mitosis and could disrupt cell surface activities necessary for the responses to mitogenic and antigenic stimuli.

This is an attractive theory, for it explains some widely varied observations about benzene cytotoxicity. First, certain benzene metabolites do react with free sulfhydryl groups. This was first reported by Tunek et al.[27] and confirmed by others.[23,28] Second, bone marrow cells from animals exposed to benzene show altered distributions of cell cycle phases. Cells from rats exposed to benzene show an increase in the number of cells in the G_2-M phases of the cell cycle and a decrease in the number of cells in the G_0-G_1 phases.[19] This would occur if cells were unable to complete mitosis and enter the G_0-G_1 phases. Third, there is evidence that only cells actively in cell cycle are affected by benzene toxicity. Noncycling lymphocyte precursors[25] and noncycling pluripotent hematopoietic stem cells[29] are apparently resistant to the toxic action of benzene or its metabolites. It appears, therefore, that benzene exerts at least some of its cytotoxicity by preventing actively cycling cells from completing mitosis. Disruption of cytoskeletal function is consistent with this mode of action.

The sensitivity of lymphocytes to benzene and their abilities to respond vigorously to mitogens have been employed to study the abilities of benzene and its metabolites to induce sister chromatid exchanges (SCE) in populations of proliferating lymphocytes.[30-33] This assay has been particularly useful in assigning relative mutagenic potencies to benzene and its metabolites. Catechol has been found to be the most potent mutagen in this assay followed by 1,4-benzoquinone, hydroquinone, 1,2,4-benzenetriol, phenol, and benzene.[32,33] Benzene without metabolic activation was only marginally mutagenic in this assay.[31-33] With the exception of phenol, these metabolites were one to two orders of magnitude more active than benzene at inducing SCE.[33] These results strengthen the hypothesis that the metabolites of benzene play a far greater role in the genotoxic effects of benzene than does free benzene.

Recently, the mitogenic responses of lymphocytes have been used to assess the toxic effects of low concentrations of inhaled benzene.[34] Short-term exposures (6 days) of mice to gradated concentrations of benzene induced depressions in the abilities of T- and B-lymphocytes to respond to mitogenic stimuli. Particularly noteworthy was the reduction in the response of B-lymphocytes after exposure to 10 ppm benzene, the current occupational exposure limit.[34] These results suggest that mitogenic assays of lymphocytes may be useful in screening individuals occupationally exposed to benzene.

B. Red Cells and Granulocytes

A discussion of the effects of benzene on granulocytes and red cells should be considered against a background of the current understanding about the normal production of these cells. A brief review of erythrocyte and granulocyte production follows and is based on several sources.[35-37]

Red cells and granulocytes arise from a common stem cell dubbed the pluripotent hematopoietic stem cell. These stem cells are believed capable of self-renewal. There is strong evidence that most (about 80%) of these cells are not actively replicating in vivo. However, they can respond to a hematopoietic insult with vigorous proliferation. Pluripotent stem cells can be enumerated by counting the number of hematopoietic colonies they form in the spleens of lethally irradiated host animals.[38] For this reason they are often called colony-forming units — spleen, or CFU-S.

It is believed that CFU-S differentiate to cells that are still capable of self-renewal but are committed to one particular hematopoietic lineage. These cells are called specific progenitor cells. They are more likely to be actively proliferating than CFU-S in vivo. Under the stimulus of an appropriate growth factor, specific progenitor cells produce hematopoietic colonies of a single lineage in vitro. They can be enumerated by counting the colonies they form in culture. Their short-hand notation is derived from their in vitro colony-forming abilities. There are two erythrocytic-specific progenitor cells. The more primitive cell is called the burst-forming unit — erythroid (BFU-E) while the more differentiated cell is called the colony-forming unit — erythroid (CFU-E). Analogously, the granulocyte/macrophage-specific progenitor cell is known as the colony-forming unit — granulocyte/macrophage (CFU-GM).

Specific progenitor cells are believed to differentiate into blasts. These are the most primitive cells recognizable by standard histologic techniques. Blasts and their immediate progeny are capable of cell division, but they usually produce daughter cells of greater maturity.

Both red cell and granulocyte precursors eventually differentiate into cells incapable of mitosis. During this process, the cells acquire the full functional capacities of mature granulocytes or red cells. At the end of this process, they are normally released into the peripheral blood.

Although blood cell differentiation probably does not occur in discreet stages, cells with common characteristics are usually grouped into compartments or pools. The stem cell pool consists of the pluripotent stem cells and the specific progenitor cells. The differentiating, proliferating pool (DPP) consists of the blasts and their progeny capable of cell division. The nonproliferating pool (NPP) consists of maturing, nondividing precursor cells, and the functional pool consists of circulating cells.

As noted previously, protracted benzene exposures induce depressions in circulating red cell numbers while circulating granulocyte levels are often markedly increased after these exposures. Benzene appears to disrupt red cell production at all the major stages of red cell development. The erythrocytic-specific progenitor cells are particularly susceptible to benzene toxicity. For example, CFU-E from mice exposed to 10 ppm benzene (the current occupational exposure limit) show marked depressions in their in vitro colony-forming abilities.[39] Moreover, the colony-forming abilities of normal BFU-E and CFU-E are inhibited when these cells are cocultured with peripheral blood mononuclear cells from rabbits exposed to benzene.[40] This last finding is particularly interesting since it indicates the sensitivity of red cell progenitors to benzene. The progenitors were not exposed to benzene, but merely cocultured with cells that were exposed, yet this was sufficient to reduce normal colony production.

More mature red cell precursors (those in the differentiating, proliferating pool) are also affected by benzene. The numbers of recognizable red cell precursors are depressed by benzene exposures[20,41] as are the abilities of these cells to incorporate iron.[42-44] The latter finding may indicate benzene-induced disruption of hemoglobin synthesis.

There is also evidence that mature circulating red cells have reduced lifespans in animals exposed to benzene. Elevated levels of splenic hemosiderin pigments have been observed in benzene-exposed animals[16-18,45] which is indicative of enhanced splenic sequestering of circulating red cells.

Thus, red cells at all stages of development that are currently assayable appear to be vulnerable to benzene. Whether the reductions in the numbers of recognizable red cell precursors are due to reduced input from the stem cell compartment alone or due to destruction of cells in the DPP as well is not known. It appears, however, that there is little or no compensatory production of red cells in the DPP to counter the effects of benzene.[20,41] For example, in one study in which mice were exposed to an exceedingly toxic concentration of benzene, the number of femoral nucleated red cells in exposed mice was 69% of the control values after 5 days and after 50 days of exposure.[20] As we shall see, granulocyte populations respond very differently to benzene.

Elevated exposures to benzene have been shown to depress granulocyte progenitor cells (CFU-GM);[46-48] however, there is evidence that these progenitor cells are less susceptible to benzene than other cells in the stem cell pool. For example, in a series of experiments in which the colony-forming abilities of both the CFU-S and the CFU-GM were determined at various benzene concentrations, the CFU-GM were more resistant than the CFU-S to benzene-induced reductions in colony formation.[46] Although direct comparisons of the effects of benzene on the colony-forming abilities of the erythroid and granulocytic-specific progenitor cells have not yet been published, there are indications that CFU-E are more sensitive to benzene than CFU-GM. Exposure of mice to 10 ppm benzene for 6 hr/day, 5 days/week, for 66 days (50 exposure days) reduced the colony-forming abilities of the CFU-E to about 40% of control values.[39] This exposure regime represents 500 ppm days of treatment. On the other hand, exposure of mice to 100 ppm benzene for 6 hr/day for 5 days, which also represents an exposure regime of 500 ppm days, caused no depression in the colony-forming abilities of the CFU-GM.[46] Although a comparison of this sort is tenuous, it does give some indication of the relative sensitivities to benzene of the CFU-E and CFU-GM.

In the DPP and the NPP granulocytic precursors show marked relative increases in number in response to benzene exposures. In the initial stages of exposure, the numbers of these cells are often depressed. However, as exposures proceed, their numbers begin to recover.[20,41] In mice at least, this recovery occurs in both the marrow and spleen.[20,49] In the marrow the increased numbers of granulocytic precursors appear to be due to increased production of myeloblasts and/or promyelocytes.[50]

Thus, granulocytic precursors respond differently than erythrocytic precursors to benzene exposures. Both show initial depressions, but the granulocytic cells increase in number during protracted exposures while the erythrocytic precursors remain depressed.[20,41] Since both cell types arise from a common stem cell, it appears that protracted benzene exposure induces a shift from a normal differentiation pattern to one in which granulopoiesis predominates. Indeed, mice killed by protracted benzene exposures often show evidence of enhanced granulopoiesis at autopsy.[17] The elevated levels of circulating granulocytes observed during prolonged benzene exposure are likely due to this shift toward granulopoiesis. The relevance of this shift in the context of the effects of benzene on pluripotent hematopoietic stem cells will now be discussed.

Benzene exposure reduces the colony-forming abilities of pluripotent hematopoietic stem cells (CFU-S).[29,46,47,51,52] The lowest reported in vivo exposure regime at which marked reduction occurs is 100 ppm × 6 hr/day × 5 days.[46] Benzene appears to disrupt normal CFU-S colony formation by injuring the cells directly and by providing an environment which will not support normal growth. In most experiments, CFU-S are removed from benzene-treated animals and then stimulated to divide by injection into lethally irradiated, syngeneic hosts.[29,46,47,52] Any diminution in colony formation must be due to benzene-induced injury of the CFU-S before transplantation. However, it has also been demonstrated that bone marrow stromal cells from benzene-treated mice do not support the growth of normal CFU-S in long-term bone marrow cultures.[51] In this latter case, only the stromal cells came from benzene-animals while the cultured CFU-S received no direct benzene exposure. Therefore, in culture at least, benzene is capable of altering the hematopoietic environment such that normal CFU-S colony formation is inhibited. A similar study in which peripheral blood cells from benzene-treated animals inhibited colony growth of the red cell progenitors has already been discussed.[40]

It is clear then that benzene treatment disrupts normal CFU-S colony formation even after relatively brief exposures. This disruption of normal hematopoietic stem cell proliferation combined with the previously discussed alteration in normal differentiating patterns are indicative of myeloproliferative disorders.[53] These disorders are related syndromes which involve the benign or malignant proliferation of hematopoietic elements.[53] Many benign manifestations of these disorders precede the onset of myelogenous leukemia or its variants. Myelogenous leukemia, the most common form of leukemia associated with benzene exposure in humans,[54] is exceedingly rare in rodents.[55] It is also an exceedingly rare occurrence in rodents exposed to benzene.[45,56,57] Yet, when rodents are given prolonged benzene exposure regimes, they exhibit signs analogous to myeloproliferative disorders. Why frank leukemia is not a more prevalent result of these exposure regimes is not known. This remains a fertile area for further study. The results of such work may bear not only on benzene toxicity, but on the study of the regulation of hematopoiesis as well.

III. CARCINOGENESIS

The association between human exposure to benzene and leukemia (predominantly, myelogenous leukemia)[54] led to many attempts to experimentally reproduce the leukemogenic effects of benzene. For the most part, these attempts met with negative results.[16,45,56-59] Rodents were the animals of choice in these studies and the rare occurrence of myelogenous leukemia in these animals, even after treatment with known leukemogenic agents,[55] probably made them poor models for the study of benzene leukemogenesis.

Although an animal model for benzene-induced myelogenous leukemia has not been developed, there has been progress in several areas concerning the carcinogenic potential of benzene. For example it has been shown that the bone marrow, the putative target organ for benzene-induced leukemia, can metabolize benzene to compounds which irreversibly bind to bone marrow macromolecules.[60-63] More importantly, it has recently been determined that bone marrow mitochondria can metabolize benzene to compounds which form adducts with bone marrow DNA.[63] The formation of such adducts is regarded as the initial step in the carcinogenic process.[64] Finally, although myelogenous leukemia has not been decisively produced in animals, the carcinogenic potential of benzene has been conclusively demonstrated in several studies.

The bioassay technique in which animals are subjected to near lifetime exposures to the highest tolerated doses of a substance and then scanned at autopsy for evidence of neoplasia, led to the experimental confirmation of the carcinogenicity of benzene. Maltoni and Scarnato[65] reported that 25% of the female rats given large doses of benzene (250 mg/kg/day × 4 to

5 days/week × 52 weeks) via oral gavage developed zymbal gland carcinomata. This result was striking for two reasons. First, the tumor yield was brisk and, second, the tumors arose at an anatomical site (inner ear canal) not previously considered a target for benzene. With these developments, the zymbal gland was particularly scrutinized in subsequent investigations of benzene carcinogenicity. In gavage studies using even higher doses of benzene, Maltoni et al.[66] confirmed their earlier findings of the production of zymbal gland tumors, this time in both male and female rats, and they also observed oral cavity carcinomata in both sexes. Zymbal gland tumors have also been observed in rats undergoing inhalation exposures to benzene.[45,66] To date, the lowest inhaled concentration that has produced zymbal gland tumors in rats has been 100 ppm.[45]

A comprehensive bioassay of the tumorigenic effects of benzene was conducted by the National Toxicology Program.[67] Groups of rats and mice of both sexes were treated with graded doses of benzene via oral gavage (25 mg/kg/day to 200 mg/kg/day × 5 days/week × 103 weeks). Male rats exhibited positive dose-related trends for zymbal gland, oral cavity, and skin carcinomata. Female rats exhibited dose-related trends for zymbal gland and oral cavity carcinomata. Male mice showed positive trends for zymbal gland, lung, and preputial gland carcinomata, and for lymphoma. Female mice showed positive responses for zymbal gland, lung, and mammary carcinomata.

The experimental induction of hematopoietic tumors by benzene appears to be limited to the lymphoid system. Two inhalation studies employing C57B1 mice exposed to 300 ppm benzene were positive for the induction of lymphoma.[17,68] One study used a 16-week exposure protocol[68] and the other a lifetime exposure protocol.[17] The 16-week protocol seems to induce a greater tumor incidence, but the lifetime exposure protocol seems to shorten the time of appearance of tumors.

Related to its carcinogenic potential, benzene also possesses clastogenic properties. In addition to producing sister chromatid exchanges, which has already been discussed, benzene also induces chromosomal abnormalities (commonly chromatid breaks and gaps)[69-71] and micronuclei in reticulocytes.[48,71,72] The latter are nuclear remnants that would ordinarily have been extruded by the developing red cell. The micronuclei assay may be a particularly sensitive measure of benzene toxicity since low doses of benzene, delivered either p.o.[71] or by inhalation,[47] induce marked increases in micronuclei.

Finally, although benzene is an animal carcinogen and clastogen, it is apparently not an animal teratogen. Even concentrations that are markedly fetotoxic or toxic to the dams do not appear to be teratogenic.[73-77]

REFERENCES

1. **Laskin, S. and Goldstein, B.,** Benzene toxicity, a critical evaluation, *J. Toxicol. Environ. Health,* Suppl. 2, 1, 1977.
2. **Withey, R. and Hall, J.,** The joint toxic action of perchloroethylene with benzene or toluene in rats, *Toxicology,* 4, 5, 1975.
3. **Haley, T.,** Evaluation of the health effects of benzene inhalation, *Clin. Toxicol.,* 11(5), 531, 1977.
4. **Ayers, G. W. and Muder, R. E.,** *Encyclopedia of Chemical Technology,* Vol. 3, 2nd ed., John Wiley & Sons, New York, 1964.
5. **Le Noir, C.,** Sur un cas de purpura attribue a l'intoxication par le benzene, *Bull. Mem. Soc. Med. Hop. Paris,* 14, 1251, 1897.
6. **Santesson, C. G.,** Uber Chronische Vergiftungen mit Steinkohlenteerbenzine; vier Todesfalle, *Arch. Hyg. Berl.,* 31, 336, 1897.
7. **Selling, L.,** Preliminary report of some cases of purpura haemorrhagica due to benzol poisoning, *Johns Hopkins Hosp. Rep.,* 21, 33, 1910.

8. **Hamilton, A.,** The growing menance of benzene (benzol) poisoning in American industry, *J. Am. Med. Assoc.,* 78(9), 627, 1922.

9. **Selling, L.,** Benzol as a leucotoxin. Studies on the degeneration and regeneration of the blood and hematopoietic organs, *Johns Hopkins Hosp. Rep.,* 27, 83, 1916.

10. **Weiskotten, H., Schwartz, S., and Steensland, H.,** The action of benzol. The deuterophase of the biphasic leucopenia and antigen-antibody reaction, *J. Med. Res.,* 35, 63, 1916.

11. **Hough, V. and Freeman, S.,** Relative toxicity of commercial benzene and a mixture of benzene, toluene, and xylene, *Fed. Proc. Fed. Am. Soc. Exp. Biol.,* 3, 20, 1944.

12. **Wolf, M., Rowe, V., McCallister, D., Hollingsworth, R., and Oyen, F.,** Toxicological studies of certain alkylated benzenes and benzene, *AMA Arch. Ind. Health,* 14, 387, 1956.

13. **Deichmann, W., MacDonald, W., and Bernal, E.,** The hemopoietic tissue toxicity of benzene vapors, *Toxicol. Appl. Pharmacol.,* 5, 201, 1963.

14. **Svirbely, J., Dunn, R., and von Oettingen, W.,** The chronic toxicity of moderate concentrations of benzene and of mixtures of benzene and its homologues for rats and dogs, *J. Ind. Hyg. Toxicol.,* 26, 37, 1944.

15. **Gerarde, H.,** Toxicological studies on hydrocarbons. II. A comparative study of the effect of benzene and certain mono-n-alkylbenzenes on hemopoiesis and bone marrow metabolism in rats, *Arch. Ind. Health,* 13, 468, 1956.

16. **Snyder, C. A., Goldstein, B., Sellakumar, A., Wohlman, S., Bromberg, I., Erlichman, M., and Laskin, S.,** Hematotoxicity of inhaled benzene to Sprague-Dawley rats and AKR mice at 300 ppm, *J. Toxicol. Environ. Health,* 4, 605, 1978.

17. **Snyder, C. A., Goldstein, B. D., Sellakumar, A. R., Bromberg, I., Laskin, S., and Albert, R. E.,** The inhalation toxicology of benzene: incidence of hematopoietic neoplasms and hematotoxicity in AKR-J and C57B1/6J mice. *Toxicol. Appl. Pharmacol.,* 54, 323, 1980.

18. **Snyder, C. A., Goldstein, B., Sellakumar, A., Bromberg, I., Laskin, S., and Albert, R. E.,** Toxicity of chronic benzene inhalation: CD-1 mice exposed to 300 ppm, *Bull. Environ. Contam. Toxicol.,* 29, 385, 1982.

19. **Irons, R., Heck, H., Moore, B., and Muirhead, K.,** Effects of short-term benzene administration on bone marrow cell cycle kinetics in the rat, *Toxicol. Appl. Pharmacol.,* 51, 399, 1979.

20. **Green, J., Snyder, C. A., LoBue, J., Goldstein, B., and Albert, R.,** Acute and chronic dose/response effect of benzene inhalation on the peripheral blood, bone marrow, and spleen cells of CD-1 male mice, *Toxicol. Appl. Pharmacol.,* 59, 204, 1981.

21. **Irons, R. and Moore, B.,** Effect of short-term benzene administration on circulating lymphocyte subpopulations in the rabbit: evidence of a selective B-lymphocyte sensitivity, *Res. Commun. Chem. Pathol. Pharmacol.,* 27(1), 147, 1980.

22. **Wierda, D., Irons, R., and Greenlee, W.,** Immunotoxicity in C57B1/6 mice exposed to benzene and Aroclor 1254, *Toxicol. Appl. Pharmacol.,* 60, 410, 1981.

23. **Pfeifer, R. and Irons, R.,** Inhibition of lectin-stimulated lymphocyte agglutination and mitogenesis by hydroquinone: reactivity with intracellular sulfhydryl groups, *Exp. Mol. Pathol.,* 35, 189, 1981.

24. **Pfeifer, R. and Irons, R.,** Effect of benzene metabolites on phytohemagglutinin-stimulated lymphopoiesis in rat bone marrow, *J. Reticuloendothelial Soc.,* 31, 155, 1982.

25. **Wierda, D. and Irons, R.,** Hydroquinone and catechol reduce the frequency of progenitor B lymphocytes in mouse spleen and bone marrow, *Immunopharmacology,* 4, 41, 1982.

26. **Irons, R. and Neptun, D.,** Effects of principal hydroxymetabolites of benzene on microtubule polymerization, *Arch. Toxicol.,* 45, 297, 1980.

27. **Tunek, A., Platt, K., Przybylski, M., and Oesch, F.,** Multi-step metabolic activation of benzene. Effect of superoxide dismutase on covalent binding to microsomal macromolecules, and identification of glutathione conjugates using high pressure liquid chromatography and field desorption mass spectrometry, *Chem. Biol. Interact.,* 33, 1, 1980.

28. **Sawahata, T. and Neal, R.,** Biotransformation of phenol to hydroquinone and catechol by rat liver microsomes, *Mol. Pharmacol.,* 23, 453, 1983.

29. **Gill, D., Jenkins, V., Kempen, R., and Ellis, S.,** The importance of pluripotential stem cells in benzene toxicity, *Toxicology,* 16, 163, 1980.

30. **Morimoto, K. and Wolff, S.,** Increase of sister chromatid exchanges and perturbations of cell division kinetics in human lymphocytes by benzene metabolites, *Cancer Res.,* 40, 1189, 1980.

31. **Morimoto, K.,** Induction of sister chromatid exchanges and cell division delays in human lymphocytes by microsomal activation of benzene, *Cancer Res.,* 43, 1330, 1983.

32. **Morimoto, K., Wolff, S., and Koizumi, A.,** Induction of sister-chromatid exchanges in human lymphocytes by microsomal activation of benzene metabolites, *Mutat. Res.,* 119, 355, 1983.

33. **Erexson, G., Wilmer, J., and Kligerman, A.,** Sister chromatid exchange induction in human lymphocytes exposed to benzene and its metabolites *in vitro, Cancer Res.,* 45, 2471, 1985.

34. **Rozen, M. G., Snyder, C. A., and Albert, R. E.**, Depression in B and T lymphocyte mitogen-induced blastogenesis in mice exposed to low concentrations of benzene, *Toxicol. Lett.*, 20, 343, 1984.

35. **Dunn, C.**, *Current Concepts in Erythropoiesis*, John Wiley & Sons, New York, 1983.

36. **Tavassoli, M. and Yoffey, J.**, *Bone Marrow: Structure and Function*, Alan R. Liss, New York, 1983.

37. **Golde, D. and Marks, P.**, *Normal and Neoplastic Hematopoiesis*, Alan R. Liss, New York, 1983.

38. **Till, J. and McCulloch, E.**, A direct measurement of radiation sensitivity of normal mouse bone marrow cells, *Radiat. Res.*, 14, 213, 1961.

39. **Baarson, K. A., Snyder, C. A., and Albert, R. E.**, Repeated exposures of C57B1 mice to 10 ppm inhaled benzene markedly depressed erythropoietic colony formation, *Toxicol. Lett.*, 20, 337, 1984.

40. **Haak, H. and Speck, B.**, Inhibition of CFU-E and BFU-E by mononuclear peripheral blood cells during chronic benzene treatment in rabbits, *Acta Haematol.*, 67, 27, 1982.

41. **Baarson, K. A., Snyder, C. A., Green, J. D., Sellakumar, A. R., Goldstein, B. D., and Albert, R. E.**, The hematotoxic effects of inhaled benzene on peripheral blood, bone marrow and spleen cells are increased by ingested ethanol, *Toxicol. Appl. Pharmacol.*, 64, 393, 1982.

42. **Lee, E., Kocsis, S., and Snyder, R.**, Dose-dependent inhibition of ^{59}Fe incorporation into erythrocytes after a single dose of benzene, *Res. Commun. Chem. Pathol. Pharmacol.*, 5, 547, 1973.

43. **Lee, E., Kocsis, J., and Snyder, R.**, Acute effects of benzene on ^{59}Fe incorporation into circulating erythrocytes, *Toxicol. Appl. Pharmacol.*, 27, 431, 1974.

44. **Andrews, L., Lee, E., Witmer, C., Kocsis, J., and Snyder, R.**, Effects of toluene on the metabolism, disposition, and hematopoietic toxicity of ^{3}H-benzene, *Biochem. Pharmacol.*, 26, 293, 1977.

45. **Snyder, C. A., Goldstein, B. D., Sellakumar, A. R., and Albert, R. E.**, Evidence for hematotoxicity and tumorigenesis in rats exposed to 100 ppm benzene, *Am. J. Ind. Med.*, 5, 429, 1984.

46. **Green, J. D., Snyder, C. A., LoBue, J., Goldstein, B. D., and Albert, R. E.**, Acute and chronic dose/response effects of inhaled benzene on multipotential hematopoietic stem (CFU-S) and granulocyte/macrophage progenitor (GM-CFU-C) cells in CD-1 mice, *Toxicol. Appl. Pharmacol.*, 58, 492, 1981.

47. **Uyeki, E., Askkar, A., Shoeman, D., and Bisel, T.**, Acute toxicity of benzene inhalation to hematopoietic precursor cells, *Toxicol. Appl. Pharmacol.*, 40, 49, 1977.

48. **Toft, K., Olofsson, T., Tunek, A., and Berlin, M.**, Toxic effects on mouse bone marrow caused by inhalation of benzene, *Arch. Toxicol.*, 51, 295, 1982.

49. **Rosenthal, G. J. and Snyder, C. A.**, The effects of ethanol and the role of the spleen in benzene induced hematotoxicity, *Toxicology*, 30, 283, 1984.

50. **Snyder, C. A., Green, J. D., LoBue, J., Goldstein, B. D., Valle, C. D., and Albert, R. E.**, Protracted benzene exposure causes a proliferation of myeloblasts and/or promyelocytes in CD-1 mice, *Bull. Environ. Contam. Toxicol.*, 27, 17, 1981.

51. **Harigaya, K., Miller, M., Cronkite, E., and Drew, R.**, The detection of *in vivo* hematotoxicity of benzene by *in vitro* liquid bone marrow cultures, *Toxicol. Appl. Pharmacol.*, 60, 346, 1981.

52. **Cronkite, E., Inoue, T., Carsten, A., Miller, M., Bullis, J., and Drew, R.**, Effects of benzene inhalation on murine pluripotent stem cells, *J. Toxicol. Environ. Health*, 9, 411, 1982.

53. **Miale, J.**, *Laboratory Medicine: Hematology*, 4th ed., C.V. Mosby, St. Louis, 1972.

54. **Goldstein, B. D.**, Hematotoxicity in humans, Benzene Toxicity, A Critical Evaluation, *J. Toxicol. Environ. Health*, Suppl. 2, 1977.

55. **Moloney, W.**, Primary granulocytic leukemia in the rat, *Cancer Res.*, 34, 3049, 1974.

56. **Goldstein, B. D., Snyder, C. A., Laskin, S., Bromberg, I., Albert, R. E., and Nelson, N.**, Myelogenous leukemia in rodents inhaling benzene, *Toxicol. Lett.*, 13, 169, 1982.

57. **Goldstein, B. D. and Snyder, C. A.**, Benzene leukemogenesis, in *Genotoxic Effects of Airborne Agents*, Tice, R. T., Costa, D. L., and Schaich, K. M., Eds., Plenum Press, New York, 1982, 277.

58. **Amiel, J.-L.**, Essai negatif d'induction de leucemies chez les souris par le benzene, *Rev. Fr. Etud. Clin. Biol.*, 5, 198, 1960.

59. **Ward, J., Weisberger, J., Yamamoto, R., Benjamin, T., Brown, C., and Weisberger, E.**, Long-term effects of benzene in C57B1/6N mice, *Arch. Environ. Health*, 30, 22, 1975.

60. **Snyder, R., Lee, E., and Kocsis, J.**, Binding of labeled benzene metabolites to mouse liver and bone marrow, *Res. Commun. Chem. Pathol. Pharmacol.*, 20(1), 191, 1978.

61. **Andrews, L., Sasame, H., and Gillette, J.**, ^{3}H-benzene metabolism in rabbit bone marrow, *Life Sci.*, 25(7), 567, 1979.

62. **Irons, R., Dent, J., Baker, T., and Rickert, D.**, Benzene is metabolized and covalently bound in bone marrow *in situ*, *Chem. Biol. Interact.*, 30, 241, 1980.

63. **Rushmore, T., Snyder, R., and Kalf, G.**, Covalent binding of benzene and its metabolites to DNA in rabbit bone marrow mitochondria *in vitro*, *Chem. Biol. Interact.*, 49, 133, 1984.

64. **Pitot, H., Grosso, L., and Dunn, T.**, The role of the cellular genome in the stages of carcinogenesis, *Genes and Cancer*, Alan R. Liss, New York, 1984.

65. **Maltoni, C. and Scarnato, C.**, First experimental demonstration of the carcinogenic effects of benzene, *Med. Lav.*, 5, 352, 1979.

66. **Maltoni, C., Conti, B., and Cotti, G.,** Benzene: a multipotential carcinogen. Results of long-term bioassays performed at the Bologna Institute of Oncology, *Am. J. Ind. Med.* 4, 589, 1983.

67. National Toxicology Program Technical Report on The Toxicology and Carcinogenesis Studies of Benzene, Draft NTP TR 289, NIH Publication No. 84-2545, NTP-84-072, 1984.

68. **Cronkite, E., Bullis, J., Inoue, T., and Drew, R.,** Benzene inhalation produces leukemia in mice, *Toxicol. Appl. Pharmacol.,* 75, 358, 1984.

69. **Meyne, J. and Legator, M.,** Sex-related differences in cytogenetic effects of benzene in the bone marrow of Swiss mice, *Environ. Mutat.,* 2, 43, 1980.

70. **Tice, R., Costa, D., and Drew, R.,** Cytogenetic effects of inhaled benzene in murine-bone marrow: induction of sister chromatid exchanges, chromosomal aberrations, and cellular proliferation inhibition in DBA/2 mice, *Proc. Natl. Acad. Sci.,* 77(4), 2148, 1980.

71. **Siou, G., Conan, L., and el Haitem, M.,** Evaluation of the clastogenic action of benzene by oral administration with two cytogenetic techniques in mouse and Chinese hamster, *Mutat. Res.,* 90, 273, 1981.

72. **Choy, W., MacGregor, J., Shelby, M., and Maronpot, R.,** Induction of micronuclei by benzene in B6C3F$_1$ mice: retrospective analysis of peripheral blood smears from NTP carcinogenesis bioassay, *Mutat. Res.,* 143, 55, 1985.

73. **Green, J., Leong, B., and Laskin, S.,** Inhaled benzene fetotoxicity in rats, *Toxicol. Appl. Pharmacol.,* 46, 9, 1978.

74. **Hudak, A. and Ungvary, G.,** Embryotoxic effects of benzene and its methyl derivatives: toluene, xylene, *Toxicology,* 11, 55, 1978.

75. **Murray, F., John, J., Rampy, L., Kuna, R., and Schwetz, B.,** Embryotoxicity of inhaled benzene in mice and rabbits, *Am. Ind. Hyg. Assoc. J.,* 40, 993, 1979.

76. **Mehlman, M., Schreiner, C., and Mackerer, C.,** Current status of benzene teratology: a brief review, *J. Environ. Pathol. Toxicol.,* 4, 123, 1980.

77. **Kuna, R. and Kapp, R.,** The embryotoxic/teratogenic potential of benzene vapor in rats, *Toxicol. Appl. Pharmacol.,* 57, 1, 1981.

Chapter 4

BENZENE METABOLISM
(TOXICOKINETICS AND THE MOLECULAR ASPECTS OF BENZENE TOXICITY)

Keith R. Cooper and Robert Snyder

TABLE OF CONTENTS

I. INTRODUCTION

Bezene has been used over the last century as an industrial chemical. It has been long recognized that chronic exposure to benzene at high levels in the workplace leads to bone marrow depression and aplastic anemia.[1] The concept that benzene was also associated with the etiology of acute myelogenous leukemia and some of its variants was slower to develop because it is a rare disease and because aplastic anemia was more readily recognized than leukemia during the early days of intense human exposure to benzene.[2] Benzene is unusual in the fact that humans appear to be more susceptible to the leukemogenic potential than most animal species. Animal models for the leukemogenic action have not been developed; however, most animals exposed to benzene develop bone marrow hypoplasia. Based on a number of epidemiology studies there is little doubt that benzene is a human carcinogen.[3,4] Several recent animal studies have shown promise for the development of an animal leukemogenic model.[5] The animal models that are being developed may result in the illucidation of the mechanism by which benzene exposure ultimately leads to leukemia. The specific topics which will be discussed in this chapter include absorption, distribution, and metabolism of benzene as they relate to the observed toxic responses of aplastic anemia and leukemia.

II. TOXICOKINETICS

A. Human Exposure To Benzene

Benzene is ubiquitous in the environment and low level exposure occurs through water, natural food stuffs, and from atmospheric exposure. It has been estimated that 20 lb of benzene is lost per ton of benzene produced during transfer and storage of benzene, with approximately 94% lost as air emissions and 6% as water effluents.[6] There is very limited data available on benzene present in water, but it has been detected in U.S. water supplies in the range of 0.1 to 0.3 ppb.[7] The contamination of water supplies from gasoline also poses a risk for human exposure to benzene because benzene is blended into gasoline and is a natural component of petroleum products. Benzene is slightly soluble in fresh water, 0.8 parts by weight in 1000 parts of water at 20°C. Benzene has a relatively low-calculated bioconcentration factor of 1.3, based on benzene octanol/water partitioning.[7] The absorption of benzene across the skin as a vapor or in aqueous media is minimal and is not therefore a major route for exposure. There are several reports in the literature on the occurrence of benzene in food stuffs.[8,9] Benzene occurs naturally in fruits, fish, vegetables, nuts, dairy products, beans, and eggs. Eggs, for example, contain between 500 to 1900 µg/kg of benzene.

The respiratory route is believed to be the major source of human exposure to benzene from the work place, gasoline vapors, and automobile emissions. Because the majority of benzene that is lost during production is believed to escape into the atmosphere, inhalation is the major route of exposure. Concentrations in the air near gas stations for benzene are reported to be between 0.3 to 2.4 ppm compared to 0.017 ppb for rural areas.[10]

Therefore, benzene is ubiquitous in the environment and human exposure to low levels of benzene is unavoidable. Because of this fact, the determination of the mechanism for toxicity and the dose response relationships for the toxicity is important to understand.

B. Absorption, Distribution, and Elimination

Benzene toxicity is most frequently caused by inhalation of benzene in ambient air. Most benzene is removed from the body in the expired air after exposure ceases. The exhalation of unchanged benzene has been studied in dogs,[11] rabbits,[12] mice,[13] and rats.[14] Schrenk[11] exposed dogs to 880 ppm of benzene by inhalation and determined that after exposure the limiting feature in respiratory excretion of unchanged benzene was the tendency for benzene to accumulate in fat. Park and Williams[12] administered ^{14}C-benzene orally and recovered approximately 46.5% of the 400 mg/kg dose in the exhaled air as unmetabolized benzene. Rickert et al.[14] reported that the excretion of unchanged benzene from the lungs of rats followed a biphasic pattern suggesting a two-compartment model for distribution and a $t_{1/2}$ of 0.7 hr. This agreed with experimental $t_{1/2}$ values for various tissues which ranged from 0.4 to 1.6 hr. Andrews et al.[13] administered benzene to mice subcutaneously and recovered 72% of an 880-mg/kg dose in the expired air. Sato et al.[15,16] reported the following tissue blood partitioning coefficients for various tissues from rabbits: liver 1.6, kidney 1.1, whole brain 1.9, lung 1.3, heart 1.4, femoral muscle 1.1, bone marrow 16.2, and retroperitoneal fat 58.5. The high tissue blood partitioning coefficient for benzene explains the accumulation within fatty tissues. Simultaneously treatment with both benzene and toluene,[13,17] or benzene and piperonyl butoxide,[18] increases the excretion of unchanged benzene. These compounds appear to act by inhibiting benzene metabolism and thereby increasing the excretion of unmetabolized residual benzene.

As stated above, the most frequent route of exposure to benzene by humans is through inhalation. Toxic effects in humans have often been attributed to combined exposure by both respiration and through the skin. For example, rotogravure workers were described as washing ink from their hands in open vats of benzene.[19] In humans the amount of benzene entering the body through the direct contact with liquid benzene can be considerable, but is less than 5% of that absorbed through the lungs in the vapor phase.[20] Although Lazarew et al.[21] claimed that benzene could be absorbed by rabbits through the skin, neither Cesaro[22] nor Conca and Maltagliati[23] could demonstrate significant cutaneous absorption in humans. Nevertheless, small amounts of benzene are absorbed by this route and may not have been detected.

Following exposure to benzene, humans, like animals, eliminate unchanged benzene in the expired air.[11-14,24,25] The elimination of unchanged benzene was quantitated in a series of studies by Nomiyama and Nomiyama[26,27] who exposed men and women to benzene at levels of 52 to 62 ppm for 4 hr and determined its respiratory disposition. A mean value of 46.9% of the benzene was taken up in these subjects; 30.2% was retained, and the remaining 16.8% was excreted as unchanged benzene in the expired air. Pharmacokinetic plots of respiratory elimination were interpreted to indicate that there were three phases to the excretion described by three rate constants. The three compartments comprise the vessel-rich tissue (VRG), receiving 75% of the cardiac output, the muscle group (MG), receiving 18% of the cardiac output, and the fat tissue (FG), which receives 5% of the cardiac output. Sato[20] reports that the saturation of each compartment is directly related to the cardiac output with VRG filled first, followed by MG and FG compartments. The elimination is inversely related to the blood flow with the FG compartment being the slowest compartment for release. There were no significant differences between men and women; however, a slower FG elimination rate was observed in female subjects. This observation can be related to the greater amount of fat tissue present in women than in men. Hunter,[19] who exposed humans to benzene at 100 ppm, detected benzene in expired air 24 hr later and suggested that it was possible to back extrapolate to the concentration of benzene in the inspired air.

Based on the above described human and animal studies, benzene is rapidly absorbed through the lungs and preferentially distributes to body fats. After exposure to benzene, the majority of the parent compound is expired through the lungs, and then there is excretion of metabolites into the urine and continued release of benzene through the lungs. While the majority of the experimental evidence suggests that the liver is the major organ involved in benzene metabolism,[13,28] the bone marrow is the target site for benzene toxicity. The hepatic metabolites are carried via the blood stream and accumulate in the bone marrow[14,29] where they may undergo further enzymatic and nonenzymatic metabolism. The distribution of benzene and its metabolites into the bone marrow can be explained by the large tissue blood partitioning coefficient, and the low vascularization of the fat tissue.

C. Metabolism

1. Metabolites

It has been known for over 100 years that benzene is converted to phenol[30] and to catechol and hydroquinone.[31] The first detailed studies of the metabolism of benzene in vivo were reported by Porteous and Williams,[32,33] and with the advent of ^{14}C-benzene, these studies were extended by Parke and Williams.[12] Rabbits were given ^{14}C-benzene (0.3 to 0.5 mℓ/kg) by oral gavage, and radioactivity was determined in expired air, tissues, and urine. The major hydroxylation product was phenol which, along with some catechol and hydroquinone, is found for the most part in urine conjugated with ethereal sulfate or glucuronic acid. Unconjugated phenol has been found in mouse and rat urine after benzene administration.[13,34,35] Parke and Williams[12] also reported on the occurrence of phenylmercapturic acid and *trans,trans*-muconic acid. The occurrence of labeled carbon dioxide and *trans,trans*-muconic acid in the expired air would indicate that an opening of the ring occurred. Based on the administered dose, 43% was detected as benzene in the expired air, 1.5% as expired CO_2, and 35% was recovered as metabolites in the urine. Between 5 to 10% was present in the feces or body tissues. Hydrolysis of the urinary metabolies resulted in 23% as phenol, 4.8% hydroquinone, and 2.2% as catechol. Extensions of this work in recent years have concentrated on metabolism in various species, on the mechanism of metabolism using in vitro techniques, and on attempting to relate benzene metabolism to its toxicity.[36-38]

Some recent work has demonstrated the formation of a ring opening product, *trans,trans*-muconal dehyde, a six-carbon alpha, and beta-unsaturated aldehyde from in vitro studies with benzene-induced rat hepatic microsomes.[38] In these same studies, the traditional metabolites were identified as phenol, catechol, hydroquinone, and muconic acid. The amount of *trans,trans*-muconaldehyde formed was approximately 12% of the phenolic fraction. The demonstration of this benzene metabolite is important in light of the fact that the same researchers have demonstrated that *trans,trans*-muconaldehyde significantly decreased red blood cell count, hematocrit, hemoglobin, and bone marrow cellularity while increasing both white blood cell count and spleen weight.[37] The mechanism for the formation of this six-carbon diene dialdehyde is believed to involve a free radical mechanism. Therefore, the current research examining the mechanism of benzene toxicity are concentrating on the polyhydroxylated and ring opening products.

Determination of benzene metabolism in humans was first evaluated as a measure of exposure. Yant et al.[39] suggested that since benzene metabolites in the urine could be detected as ethereal sulfates, it would be possible to estimate benzene exposure by measuring the ratio of inorganic to organic sulfate. Normally the inorganic sulfate is present at about four times the organic levels. Exposure tends to increase the organic sulfate, and lower the inorganic. Hammond and Herman[40] suggested that of total sulfates, inorganic sulfates of 80 to 95% were normal, 70 to 80% indicated some exposure to benzene, 60 to 70% suggested a dangerous level of benzene exposure, and 0 to 60% indicated that benzene levels were sufficiently high enough to provide an extremely dangerous atmosphere for humans. In humans, the sulfate is the major conjugate of phenol until levels of approximately 400 mg/ℓ are reached. Beyond that level glucuronides are seen.[24,25] Teisinger and Skramovsky[41] exposed humans to benzene at 100 ppm for 5 hr and found that the urine contained primarily phenol with small amounts of catechol.

Based on a number of studies, it is generally accepted that benzene must undergo metabolic activation to produce its toxicity.[13,28,42,43] The two major organ systems that have been examined for benzene metabolism are the liver and, to a lesser extent, the bone marrow. The liver has been extensively examined because of the high level of mixed function oxidase (MFO) activity and because of a set of experiments which demonstrated that partial hepatectomy completely protected against benzene hematotoxicity and decreased whole body benzene metabolism by 70%.[28] Coadministration in vivo of toluene, a competitive inhibitor

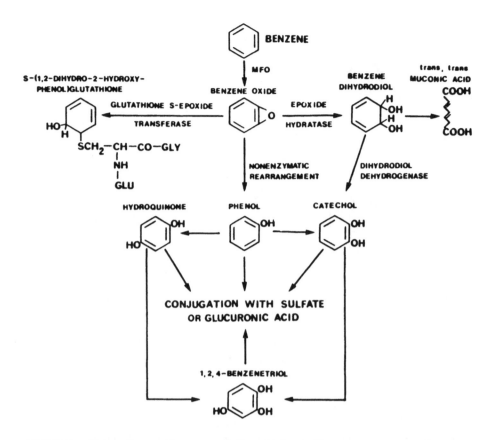

FIGURE 1.   Major pathways of benzene metabolism with the specific enzyme or nonenzymatic process involved at each metabolic step and the resultant conjugated products.

of benzene metabolism, caused a reduced urinary excretion of benzene metabolites, a concurrent reduction in benzene metabolites in bone marrow, blood, liver, spleen, and fat tissue, and controlled levels for ^{59}Fe red blood cell incorporation. Therefore, the toxic effects which are observed following benzene exposure are due to one or more of its metabolites.[42,44]

The metabolites that have been isolated from a number of different animals and man are shown in Figure 1. Phenol, hydroquinone, and catechol are the major metabolites of benzene in both man and animals. Other metabolites which have been reported include 1,3,5-trihydroxybenzene, *trans,trans*-muconic acid, *trans,trans*-muconaldehyde, and 1,2-dehydro-1,2 dihydroxy benzene. These metabolites have been isolated either as the free compound, ethereal sulfates, glucuronides, or a phenylmercapturic acid.[1,24,25] It should be noted that in humans the major conjugates below 400 mg/ℓ of benzene are ethereal sulfates.[24,25]

2. Mechanism for Benzene Metabolite Formation

Benzene is metabolized to phenol, catechol, hydroquinone, trihydroxybenzene, and several other minor metabolites.[12,30,38] Benzene may be oxidized to benzene oxide, which can rearrange to form phenol,[44,45] or benzene may be hydroxylated via a cytochrome P-450-generated hydroxyl free radical.[46] Cavalieri et al.[48] have suggested that this pathway is unlikely because benzene ionization potential does not favor direct free radical insertion. Figure 2 illustrates how cytochrome P-450 can act either by epoxide formation or free radical insertion.

Griffiths et al.[49,50] suggested (Figure 3) that benzene reacts with cytochrome P-450 to yield benzene epoxide which rearranges to form phenol; the system is not completely coupled

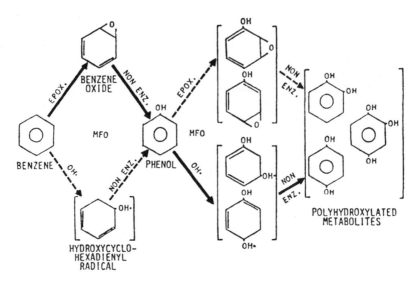

FIGURE 2.   Formation of phenol and polyhydroxylated benzenes by cytochrome P-450 (MFO) or free radical insertion. Thickened arrows indicate the predominant pathway.

resulting in periodic NADPH oxidase activity which gives rise to free radical active oxygen which in turn acts to hydroxylate phenol. The nature of the radical is unknown, but it is suggested that the hydroxyl radical is a likely candidate for phenol hydroxylation whereas superoxide is perhaps more likely to be involved in benzene epoxide formation. By the same token, the formation of phenol epoxide as an intermediate in the formation of hydroquinone or catechol is unlikely, but has not yet been completely ruled out. Additional studies are needed to clarify these points. However, this concept is compatible with both a radical mechanism and epoxide formation occurring side by side, the first step resulting in mainly epoxide formation and generation of oxygen radicals, with subsequent hydroxylations occurring mainly via radical insertion.

The subsequent metabolism of phenol in liver microsomes results in the formation of hydroquinone and catechol.[51-53] Tunek et al.[54] argued that a free radical attack on hydroquinone yielded a quinone or semiquinone capable of covalently binding and Lunte and Kissinger[55] suggested that the conversion of hydroquinone to p-benzoquinone was an enzymatic reaction. Wallin et al.[56] recently reported that metabolic activation of ^{14}C phenol in rat liver microsomes and the subsequent covalent binding was NADPH dependent. The P-450 isoenzymes catalyzing this reaction are not the major purified forms from either phenobarbital or beta-napthaflavone-induced animals. Once the phenol is metabolized to either hydroquinone or catechol, it would appear that oxidized products (not involving cytochrome P-450) covalently bind to proteins. This observation can explain the extensive covalent binding that is observed in organs which have low activity of cytochrome P-450 (i.e., spleen), but which have inorganic and organic compounds which can further oxidize these products. They also reported that covalent binding by both catechol and hydroquinone was not stimulated by NADPH and was completely inhibited by ascorbic acid (1 mM) and glutathione (10 mM).

This type of combined cytochrome P-450 and further oxidation of polyhydroxylated benzenes and subsequent covalent binding can explain the occurrence of high levels of binding in vivo in organ systems with low cytochrome P-450 activity.

It is not clear which enzymes are responsible for the metabolic activation of benzene to reactive species. There are at least two different hepatic microsomal cytochromes P-450 which hydroxylate benzene.[57-59] There appear to be enzymes capable of benzene metabolism

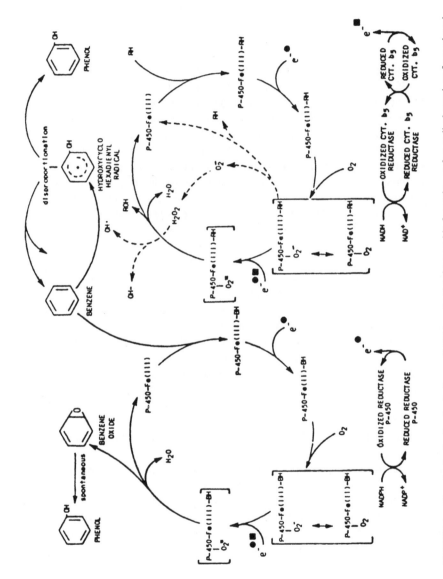

FIGURE 3. Hypothesized role of cytochrome P-450 to act either by epoxide formation or free radical attack to form phenol from benzene.

in rabbit bone marrow microsomes.[60,61] In a series of studies, it has been demonstrated that rabbit bone marrow has MFO activity as determined by aryl hydrocarbon hydroxylase activity and can metabolize benzene to phenol and an unidentified metabolite in vitro.[13] Irons et al.[62] showed that metabolism and covalent binding occurred in Fisher 344 rat bone marrow perfused with ^{14}C-benzene. They reported finding phenol, catechol, hydroquinone, and unknown metabolites covalently bound to bone marrow proteins. Benzene metabolism also occurs in rat hepatic mitochondria and in mitochondria of rat, rabbit, and cat bone marrow.[63,64] Of particular importance is the apparent association between the metabolism of benzene in mitochondria and the inhibition of RNA and protein synthesis.

Although the initial step in benzene metabolism in the liver, and perhaps the bone marrow as well, may be mediated by cytochrome P-450, the subsequent metabolism of phenol and other enzyme metabolites may result from several enzymatic and nonenzymatic processes. Some of those likely to be of significance will be discussed below.

The formation of polyhydroxylated benzene metabolites may be due to peroxidative mechanisms. The enzyme myeloperoxidase (MYOP), which is a constitutent of several types of leukocytes, has been studied for several decades. Peroxidatic activity in these cells may largely be the result of the activity of this enzyme. The concentration of H_2O_2 available for peroxidatic and other reactions is normally limited by catalase, an enzyme in the cell specifically concerned with the inactivation of hydrogen peroxide. H_2O_2, which escapes destruction by catalase, is available for use by MYOP to oxidize a variety of substrates. However, it is also clear that catalase, itself, can demonstrate a peroxidatic function under some circumstances and may contribute to the formation of reactive metabolites.[65]

Peroxidases in general and MYOP in particular have been the subjects of biochemical investigation for many years with some pertinent studies going as far back as 1855. Catalases and peroxidases are enzymes from various plant and animal sources which mediate reactions in which H_2O_2 is converted to water. In the catalase reaction, two molecules of H_2O_2 are converted to two of water and one of oxygen.[66] The peroxidase reaction, however, requires a source of reducing equivalents to modify the peroxide. Many such sources of reducing equivalents have been used which include the phenolic metabolites of benzene. However, it is likely that in leukocytes, chloride acts as the electron donor and the resulting hypochlorite production plays a significant role in the normal bacteriocidal function of these cells.

The fundamentals of the kinetics of peroxidase reactions, developed using horseradish peroxidase (HRP), were laid down in the early 1950s and were best reviewed by Chance and Maehly.[66] Agner[67,68] recommended an assay utilizing uric acid as the hydrogen donor, but also reported that phenol, catechol, and resorcinol, but not hydroquinone, were active in the system.

Most oxidations catalyzed by peroxidase involve two successive one-electron transfers. The first step involves formation of a complex with the enzyme and the peroxide, followed by successive oxidations of 2 mol of the chemical. It now appears that the peroxidase found in leukocytes, termed MYOP, is a different enzyme from peroxidases found in other cells and tissues such as eosinophils, uterus and mammary tumors of the rat, and bovine lactoperoxidase.[69] Myeloperoxidase is a distinctly different enzyme from both lactoperoxidase[70] and the peroxidase of the eosinophil.[71] It is also apparent that MYOP itself is not a single enzyme, but the name represents a family of isozymes, each of which exhibits subunit structure.[73]

Given this abbreviated review of some of the essential features of MYOP, the following relates to the implication that this enzyme system plays a role in benzene toxicity. HRP can oxidize phenol.[74] Examination of the products of phenol oxidation by HRP in these studies revealed the presence of o,o'-biphenol and p,p'-biphenol and the oxidation product p,p'diphenoquinone. Although these compounds are products of HRP and may be products of myeloperodiase activity upon phenol, they are not products seen in vivo during benzene or phenol metabolism.

Smart and Zannoni[75] demonstrated that covalent binding of phenol to microsomal protein was enhanced by the addition of a homogenate of bone marrow leukocytes rich in MYOP to guinea pig liver microsomes. They studied the metabolism of phenol by measuring the extent of covalent binding of phenol metabolites, hydroquinone, and catechol to guinea pig hepatic microsomal protein. Covalent binding was inhibited by an NADPH requiring guinea pig DT-diaphorase and an NADH requiring diaphorase from *Cana kluyveri*. The effect was prevented by the DT-diaphorase inhibitor, dicumarol. In the presence of HRP and guinea pig bone marrow, MYOP covalent binding increased 30 and 6 times, respectively. It is suggested that the selective imbalance between oxidation of benzene metabolites by MYOP and their reduction by diaphorase may account for the selective toxicity of benzene in bone marrow, whereas detoxication mechanisms may predominate in liver and thereby explain why benzene is not hepatotoxic. Smart and Zannoni,[76] in a later paper, demonstrated that when either benzene or phenol was used as the substrate for in vitro microsomal metabolism studies, ascorbate and DT-diaphorase had significantly different effects on the amount of covalent binding. Glutathione decreased the amount of covalent binding by 95%. Ascorbate added to the incubation mixture reduced covalent binding of benzene only by 71% and DT-diaphorase by only 18%. Ascorbate added to the phenol incubation mixture reduced the covalent binding of phenol by 95% and DT-diapherase was 70% inhibited. The metabolism of benzene was reduced by 35% by the addition of ascorbate, while there was no effect on phenolic metabolism. This set of experiments points up the complex nature of the role of the various metabolizing systems on both the parent substrate and the metabolites, themselves. Therefore, it is unlikely that a single enzyme system will be found to metabolize benzene to its reactive metabolites.

O'Brien[77] reported that the addition of HRP to rat liver microsomes increased the covalent binding of labeled phenol to DNA. In the work of Smart and Zannoni[75,76] and O'Brien[77] the effects of isolated, soluble, added enzymes were compared with membrane-bound mixed function oxidase. It is tempting to draw the conclusion that the peroxidases play a more significant role than the mixed function oxidase in the metabolism of benzene and its phenolic metabolites in bone marrow based on these data. For the past few decades, those of us interested in the relationship between the metabolism of xenobiotics and mechanisms of toxicity have tended to make use of liver microsomal preparations as if they were either soluble enzymes or acted in the test tube as they do in cells. However, there are no microsomes in intact cells. Microsomes are artifacts produced by the disruption of the endoplasmic reticulum during the process of homogenization. The approach of substrates to the enzymes of the endoplasmic reticulum and the fate of the metabolites must in some way be related to the structural features of the cells in which metabolism occurs. Thus, to argue that the formation of reactive metabolites formed by cytochrome P-450 in liver microsomes in vitro, and the low level of DNA adduct formation in that system is a good model for the potential for covalent binding of reactive metabolites from cytochrome P-450-mediated reactions to DNA in the cell is inaccurate. To then superimpose upon this scheme an isolated, soluble enzyme such as a peroxidase, and to provide it with adequate amounts of H_2O_2 to produce reactive metabolites from phenol is to artificially drive the reaction in a specific direction which may or may not mimic what is happening in the cell and bone marrow.

Another mechanism for the production of reactive metabolites may be the release of active oxygen during the phenomenon called the respiratory or oxidative burst. Unlike MYOP, which is always present in the cells and would generate peroxidatic activity whenever peroxide and a reducing substance is available, this process would be episodic, i.e., only occur when activated cells are stimulated to produce the respiratory burst. A variety of leukocytes can produce oxidative bursts, but the result would be the same in terms of production of reactive metabolites of benzene since the burst yields several types of active oxygen. Thus, there may be differences in reactive metabolites formed during the oxidative burst vs. the MYOP reaction.

O'Brien[77] reported that during the oxidative burst in leukocytes activated by the well-known tumor promoter, phorbol myristate acetate (PMA), phenol, the first metabolite of benzene was oxidized to a reactive metabolite which bound to DNA. The respiratory burst is a phenomenon observed in several types of white cells including neutrophils, eosinophils, monocytes, and macrophages, and is associated with either phagocytosis or a variety of other stimuli.[78-82] Ordinarily the respiratory burst occurs only in cells which are termed "activated". The respiratory burst is characterized by the sudden increase in oxygen uptake by the cells and the rapid formation of superoxide radical anion, O_2^-, hydrogen peroxide, and other reactive oxygen compounds. The cells undergoing the respiratory burst are ordinarily protected from its effects by enzymes such as superoxide dismutase, catalase, and glutathione peroxidase, but these cells may, under some conditions, be damaged or killed by a failure to protect themselves from the products of the respiratory burst. There is also the potential for cells in the immediate vicinity to be damaged by active oxygen and other factors released during the burst.

With respect to benzene toxicity, the net effect of the respiratory burst may be to produce toxic metabolites. Thus, in theory, benzene could be hydroxylated by reactive oxygen resulting from the burst or, if benzene is metabolized elsewhere, i.e., the liver or other bone marrow cells, the resulting phenol would clearly be a candidate for further hydroxylation by the radicals released during the burst.

A number of agents are thought to exert their toxic effects via oxygen reduction leading to the production of superoxide, hydrogen peroxide, hydroxyl radical, and perhaps singlet oxygen.[83] These reactions may be enzymatic or nonenzymatic. NADPH oxidase is an enzyme activity in which the mixed function oxidase reduces oxygen making use of reducing equivalents from NADPH. This may happen in the absence of a substrate or may be an ongoing activity of some types of cytochrome P-450 not involved in metabolizing the specific substrate undergoing metabolism. Lorentzen et al.[84] demonstrated that 6-hydroxybenzo[a]pyrene in solution autoxidizes to form 6,12-, 1,6-, and 3,6-benzo[a]pyrene diones in the absence of any enzymes. During the formation of these metabolites, oxygen was consumed and H_2O_2 was generated. Lorentzen et al.[85,86] showed that these metabolites were toxic to cells in culture in the presence of oxygen. The relationship of these studies to benzene metabolism and toxicity lies in the necessity for benzene to be metabolized as far as the stage of dihydroxylated metabolites, especially catechol, before autoxidation can play a role.

Therefore, the current thinking on the metabolism of benzene is that benzene is metabolized by cytochrome P-450 to yield phenol which can either be further metabolized to polyhydroxylated products by cytochrome P-450 or by reactive oxygen species. The reactive oxygen species can be generated by a number of enzymatic and nonenzymatic systems which can yield polyhydroxylated and ring opening products. The following section will discuss the relationship between specific metabolites and their toxic effects.

III. TOXIC RESPONSES

Benzene exposure in both man and experimental animals results in a generalized bone marrow depression which manifests itself as reduced numbers of circulating erythrocytes, granulocytes, thrombocytes, lymphocytes, and monocytes.[2,87,88] There is a causal relationship between benzene exposure and the occurrence of leukemia in man.[3,4] The type of leukemia most commonly associated with benzene exposure in man is acute myelogenous leukemia and the variants which include erythroleukemia and acute myelomonocytes leukemia.[89] Whether aplastic anemia is a prerequisite for leukemia is not clearly understood. With this brief introduction we will examine benzene toxicity on differentiating cells in the bone marrow, specific target sites within cells, and finally, present work on animal models for carcinogenicity and how this relates to the mechanism of benzene toxicity.

A. Affects On Specific Cells

The bone marrow is a complex matrix of rapidly dividing cells which depend on the integrity of the microenvironment, growth-regulating factors that are released from several cell types, and the capability to undergo division. The complexity of the bone marrow and the number of different parameters that have been shown to be affected by benzene and its metabolites demonstrate why there may be multiple causes for the occurrence of the aplastic anemia and leukemia.

The early studies on the mechanism of benzene-induced bone marrow depression evaluated bone marrow and for the most part emphasized morphology. The classical studies of Selling[90] included a description of rabbit bone marrow following chronic intoxication with subcutaneously administered benzene. The first signs of bone marrow damage were evident on the second day of injection with 1 mℓ/kg and by the ninth day, aplasia was complete. All of the cell types were damaged and gradually disappeared during the course of treatment. Considerable evidence has developed to support the concept that benzene produces its effect by inhibiting cell division.[91-100] Fewer mitotic figures were observed in the marrow of benzene-intoxicated animals and, in addition, benzene has been shown to cause abnormal mitotic figures[101] and decreased 3H-thymidine incorporation. Sammet et al.[28] demonstrated that livers of partially hepatectomized rats failed to grow back when the animals were treated with benzene, and D'Souza et al.[102] reported that the remaining ovary in the hemi-spayed rat did not undergo compensatory cell proliferation after treatment with benzene. These data suggest that benzene inhibited cell proliferation. Thus, it may be that the reason that benzene does not act as a primary liver toxin, but is highly effective against the bone marrow, is that liver cells do not normally undergo rapid cell proliferation whereas rapid proliferation is a property of bone marrow cells.

Several investigators examining different developing cell types in the bone marrow have shown that benzene is preferentially toxic to progenitor cells of intermediate differentiation.[91-93] The destruction of mature nondividing cells has not been observed, but effects on their capability to release various factors involved in blood cell differentiation have been observed.

Studies on the erythroid line using ^{59}Fe incorporation in mice, demonstrated that early erythroblasts are more sensitive to benzene than stem cells, reticulocytes, or mature erythrocytes.[96,103,104] It should be noted that none of the benzene metabolites including phenol, di- and tri-hydroxy benzenes, benzoquinone, 2,5-dehydroxy-1,4-benzoquinone, or *trans,trans*-muconaldehyde resulted in a depression if ^{59}Fe uptake into erythrocytes was seen with benzene administration.[96] Although a more recent report has shown significant decreases with benzene, phenol, hydroquinone, and catechol. Treatment with 3-amino-1,2,4-triazole in vivo, an inhibitor of microsomal metabolism, protected the animal against the inhibition of ^{59}Fe into RBCs produced by benzene and phenol. However, 3-amino-1,2,4-triazole had no effect on catechol or hydroquinone-induced inhibition of ^{59}Fe uptake.[36]

The most significant feature of the bone marrow is its elaborate morphology. There are large numbers of cells of many types, indicative of its role as a site of maturation and cell proliferation. In addition to stem cells and the various forms of maturing myeloid and erythroid cells, the marrow contains populations of cells that make up the "microenvironment".[91,97,105] Allen and Dexter[105] suggest that the stromal cells of the bone marrow do not simply supply a supportive and permissive millieu where developing cells can take advantage of externally derived hormonal influences which foster their maturation. They have studied a model of the reconstituted microenvironment in liquid marrow cultures and have investigated the adipocyte, a cell which they believe arises via the accumulation of lipid in the reticulum cell which is closely associated with, and may play a role in, the development of granulocytes. They also report on a group of cells termed "blanket cells" which spread out and cover tightly packed groups of granulocytes and macrophages. In cultures, macrophages and granulocyte precursors lie under the blanket cells in close proximity and the precursors begin to undergo mitosis after which they migrate out from under the blanket and form associations with the developing fat cells. The authors postulate that the macrophages play a vital nurturing role in the development of erythroid and granulocytic cells by providing essential growth factors.

The concept that benzene-induced aplastic anemia results from an effect of benzene on bone marrow stem cells is most strongly supported by studies of colony-forming units (CFU). Thus, exposure to benzene has been shown to inhibit spleen colony formation,[91,92,106] CFU-C,[93] and GM-CFU-C.[92] On the other hand, a number of in vivo observations[94,96,107,108] suggest that early maturing cells in the marrow may be important targets. Frash et al.[97] and Harigaya et al.[91] reported that benzene appears to inhibit normal activity of stromal microenvironmental cells in vivo and in vitro, respectively. Which of these cells are the most critical targets for benzene toxicity has yet to be demonstrated.

Cronkite et al.[100] studied the effects of inhaled benzene on pluripotent bone marrow stem cells using the spleen colony-forming unit technique. Spleen colony-forming units (CFU-S) result when bone marrow cells of donor mice are administered into irradiated recipient mice. The bone marrow cells from the donor mice migrate to the spleen of the recipient mice and form colonies of cells which can be observed on the surface of the spleen. The number of such colonies are interpreted to indicate the number of stem cells from the donor animals that were administered to the irradiated recipient animals. Male BNL donor mice were exposed to 400 ppm benzene for 6 hr/day, 5 days/week, for up to $9^{1}/_{2}$ weeks. Hematotoxicity was analyzed at various times after benzene exposure. Exposure to benzene caused a significant decrease in both red and white peripheral blood cell counts throughout the exposure period. A significant decrease in absolute marrow cellularity (threefold) and CFU-S content (fivefold) was observed after 5 days exposure to benzene. The number of CFU-S undergoing DNA synthesis was also decreased twofold. These authors explain the substantial decrease in absolute marrow cellularity by a large reduction in the amplifying populations of identifiable erythrocytic and granulocytic precursors.

Exposure of male C57B16 mice to 10 ppm benzene for 6 hr/day, 5 days/week caused a significant, progressive depression of in vitro colony-forming ability of one of the erythroid (E) progenitor cells, as measured by a CFU assay referred to as CFU-E.[109] Hematopoietic measurements were made on days 32, 66, and 178 of benzene exposure. Benzene exposure produced a decrease in splenic nucleated red cells and circulating red blood cells and lymphocytes. In addition, bone marrow and splenic progenitor cells from benzene-exposed mice showed a significant reduction in their ability to form colonies, compared to cells from control mice at all time points examined. The numbers of CFU-E colonies from benzene-exposed mice declined progressively during the exposure period reaching 5% of control for marrow cells and 10% of control for spleen after 178 days. Benzene-exposed mice also exhibited a significant decrease in the numbers of differentiated hematopoietic cells including circulating red blood cells and lymphocytes. Red cells were depressed at 66 and 178 days, while peripheral blood lymphocytes were depressed at all times. Benzene exposure appeared to have no effect on peripheral blood neutrophils. The authors suggested that the effects were due to a progressive inability of CFU-E to replicate and differentiate in culture. They concluded that exposures to levels of benzene considered occupationally safe could markedly disrupt normal erythropoiesis in mice.[109,110]

Rozen et al.[111] and Gram et al.[112] reported a significant depression in blastogenesis of both B and T lymphocytes following short-term exposure to inhaled benzene. Male C57B1/6J mice (7 to 8 mice per group) were exposed to benzene (nominal concentrations were 0, 10, 30, 100, or 300 ppm) for 6 hr/day, for 6 days. Peripheral blood samples were examined 30 to 90 min following the last exposure. All concentrations of benzene tested produced significantly decreased lipopolysaccharide (LPS)-induced B cell mitogenesis. At benzene levels of 31 ppm (actual measured value) or greater, there was also a significant decrease in phytohemaglutinin (PHA)-induced splenic mitogenesis. Benzene had no effect on the number of splenic T cells. Peripheral blood lymphocyte counts were depressed at all benzene levels, while red blood cell counts were depressed only at the highest benzene levels (100 and 300 ppm). The results of these studies suggested to the authors that short-term exposure to inhaled benzene at relatively low exposure levels could cause significant decreases in both lymphocyte number and function.

Irons et al.[62] reported that the most sensitive cells to benzene were peripheral lymphocytes and differentiating bone marrow precursor cells. There was a decrease in the number of bone marrow cells undergoing division, and there was an increase in the number of cells in G_2 or M phase. Bone marrow smears showed no cytotoxic effects on blast forms and mature nondividing cells. Irons and co-workers demonstrated that hydroquinone and p-benzoquinone inhibited lectin-stimulated blastogenesis in cultured lymphocytes.[113] Pfeifer and Irons[114] have shown earlier that hydroquinone, p-benzoquinone, and 1,2,4-benzenetriol, inhibited polymerization of tubulin by interacting with the sulfhydryl groups. Phenol has been shown to produce no evidence of immunotoxicity; however, both hydroquinone and catechol were immuno-suppressive. Hydroquinone reduced the cellularity of both the spleen and bone marrow, while catechol only reduced splenic cellularity. These investigators have suggested that benzene toxicity may arise through the reaction of the polyhydroxyl benzene metabolites with reactive SH groups on microtubules, thereby inhibiting their polymerization and subsequent lymphocyte mitogenesis and agglutination. Although the inhibition of the tubulin formation would result in a decreased number of cell division and the aplastic anemia, it does not account for the leukemogenic effect of benzene. The leukemogenic effect may be explained by altering growth factors released by these cells or by the transformation of cells by chemical interaction with DNA.

Nevertheless, an understanding of the relationship between lymphocytes and hemopoietic cells is of significant clinical importance to an understanding of aplastic anemia and its role in leukemia.

T-lymphocytes, both in vivo and in vitro, are a potent source of factors that regulate hemopoiesis.[115-117] A murine T-lymphocyte clone, L-2, after stimulation by concanavalin A, secretes at least two distinct glycoprotein factors, interleukin-3 (Multi-CSA IL-3) and granulocyte/macrophage colony stimulating activity (GM-CSA), that affect hemopoietic precursor cells.[117] IL-3 stimulates the formation of colonies of neutrophils, macrophages, megakaryocytes, and basophil/mast cells, has burst promoting activity, and increases the number of spleen colonies (CFU-S), which represent pluripotent stem cells. Thus, IL-3 may be a growth factor for early cells common to both the lymphoid and hemopoietic lineages.[118]

Furthermore, peripheral lymphocytopenia is an early manifestation of benzene toxicity in both animals and humans and is a distinctive feature of benzene-induced aplastic anemia.[62,92,119,120] In a benzene inhalation study, Rozen et al.[111] demonstrated that mitogen-induced blastogenesis of both T- and B-lymphocytes in mice was significantly depressed after exposure to concentrations of benzene as low as 10 ppm, the current standard for occupational exposure. Pfeifer and Irons[114] demonstrated that cell populations enriched in lymphoid precursors were inhibited from undergoing a proliferative response to mitogens after exposure to concentrations (10 ppm) of the benzene metabolites, p-benzoquinone or 1,2,4-benzenetriol.

Toxicity to bone marrow and lymphoid organs correlates with the concentrations of hydroquinone and catechol accumulated in those tissues. Hydroquinone inhibits lectin-stimulated lymphocyte activation in culture and interferes with microtubule assembly.[114,122] Suppression of lymphocyte activation by hydroquinone has been postulated to be mediated through interference with the cytoskeleton, specifically by interaction of p-benzoquinone, the oxidation product of hydroquinone, with intracellular sulfhydryl groups essential for lymphocyte agglutination and blastogenesis, and with similar groups responsible for maintaining a functional microtubule apparatus.[98] While this represents an intriguing explanation for benzene toxicity, at least in the lymphocyte, the question remains whether it can be extrapolated to other marrow cell types. It is of interest that no cases of aplastic anemia have been reported following the use of colchicine or vinblastine which also are potent tubulin inhibitors.

Benzene is also toxic to the bone marrow microenvironment of which the macrophage is an essential cell.[91,97] In lethally irradiated mice treated with benzene, injection of normal bone marrow cells could not reconstitute normal hematopoiesis in that the number of granulocyte and macrophage colonies was decreased. Harigaya et al.[91] showed that the adherent layer from benzene-treated mice had a reduced capacity to support the growth of bone marrow stem cells from normal animals. The development of the adherent layer was altered by benzene in that cells containing fat which normally appear during growth, failed to develop when marrow from benzene-treated animals was used.[123] Hydroquinone appears to inhibit the ability of the adherent layer to support the formation of CFU-GM.[124] Macrophages from benzene-treated rabbits inhibit the formation of CFU-E and BFU-E colonies when cocultured with normal bone marrow.[125] Thus, an additional site of benzene toxicity may be the bone marrow stroma and the macrophage may be a target.

A number of studies have been carried out on the mitochondria as a target organelle and as a model system for evaluating benzene toxicity.[126,127] Several interesting findings have been reported and may directly relate to the previously described effects on bone marrow differentiating cells. Inhibition of RNA synthesis in mitochondria has been demonstrated from mitochondria isolated from liver and rabbit bone marrow mitoplasts.[126] Benzoquinone and hydroquinone covalently bound to the mitochondrial DNA.

Benzene was also shown to inhibit mitochondrial translation, which may be due to a lack of messenger RNA and a subsequent disaggregation of polysomes. Both hydroquinone and its oxidative product, p-benzoquinone, inhibited DNA replication in rat liver and bone marrow mitochondria (MT) in a dose-dependent manner.[126] p-Benzoquinone, hydroquinone, and 1,2,4-benzenetriol inhibited the activity of partially purified MTDNA polymerase gamma. The interaction of the chemicals was with the $-SH$ group on the enzyme and not the template.[127]

Messenger RNA synthesis in nuclei from bone marrow cells has been shown to be inhibited by hydroquinone and p-benzoquinone with 50% inhibition at 6×10^{-6} M. Catechol and 1,2,4 benzenetriol have 50% inhibition at 10^{-4} M. Phenol did not inhibit MRNA synthesis at 10^{-3} M.[128] This type of inhibition could also explain the occurrence of aplastic anemia.

B. Interaction With DNA

The study of biological reactive intermediates has yielded many observations which indicate that metabolic activation can yield chemically reactive species which react nonenzymatically with cellular macromolecules to modify their activity and thereby produce toxicity. Symposiums on biological reactive intermediates have permitted the collection of considerable data on these reactions and their effects.[129,130] Covalent binding of reactive metabolites to DNA is thought to be a potential cause of cell pathology or carcinogenesis. We have focused on the covalent binding of benzene metabolites to DNA because benzene inhibits DNA synthesis,[95] benzene binds to DNA liver[131] and bone marrow,[126] and because the target organ, i.e., the bone marrow, has as its main concern the rapid proliferation of many cell types. Clearly, binding to DNA could have a profound effect on rapidly dividing cells.

Based on the studies cited above which showed that hydroquinone and benzoquinone formed adducts with mitochondrial DNA, studies were conducted to characterize the DNA adducts with eukaryotic DNA.[131] A direct nonenzymatic reaction of ^{14}C hydroquinone with calf thymus DNA was demonstrated. When hydroquinone is used in these reactions, a mild oxidizing agent such as iron must be added for an adduct to be formed. The NMR analysis of the deoxyguanosine adduct in D_2O using a 380MH NMR is shown in Figure 4A, and the mass spectrum of the same material is shown in Figure 4B. Based on these results, the structures and mechanism for the adduct formation are shown in Figure 5. Reacting ^{14}C hydroquinone with calf thymus DNA and *Micrococcus lysodeiktus* DNA resulted in four major labeled nucleosides derived from calf thymus DNA, and three from the bacterial DNA. These findings, as well as those which have shown a direct correlation between DNA binding and functional defects, are important in view of the evidence suggesting the effects of the polyhydroxylated benzenes on bone marrow cells. That DNA adduct formation may play a role is not in contrast with the thoughts of many investigators who have postulated the covalent binding of many xenobiotics to DNA as part of the explanation for various types of cell toxicity or the initiation of a carcinogenic response.[129,130]

Benzene was found to have no mutagenic potential when tested in *Drosophila melanogaster*.[133] In this experiment, newly hatched larvae were exposed to media containing benzene at a concentration of 1 or 2%. The test system used a stable X-chromosome ($zDpw^{+61e19}$) as a control and a genetically unstable sex-linked genotype, $sczw^+$. Mutation in the two systems is measured by a change in eye pigmentation; however, neither concentration of benzene showed mutagenic activity.

In a review article, Dean[134] reported that benzene was tested for mutagenic activity at concentrations of 20 and 60 $\mu\ell$/plate with five *Salmonella typhimurium* tester strains (TA100, TA98, TA1535, TA1537, and TA1538) with and without 9000 g supernatant fractions from liver microsomes. No mutations were observed. Furthermore, higher levels of benzene (0.1 to 1.0 mℓ/plate also showed no mutagenic effect when *S. typhimurium* tester strains TA98 and TA100 were used with the addition of 9000 g supernatant fractions from either phenobarbital or 3-methylcholanthrene-treated rats. Preincubation of benzene with the metabolic activating system also produced negative results. In addition, in this same set of experiments, a host-mediated assay was performed in which mice were first treated with phenobarbital and then given two 0.1 mℓ subcutaneous injections of benzene. The test organism was *S. typhimurium* strain TA1950. Again, no increase in mutation rate was observed. It is possible that any mutagenicity of benzene cannot be assessed with bacterial systems because benzene may damage the cell walls. Toluene is routinely used to render the membranes of *Escherichia coli* permeable to nucleotides while allowing the cells to maintain both their structure and their ability to synthesize DNA. However, these bacteria are not viable and changes in the membrane are difficult to detect by bright-field microscopy.[134] Pulkrabek et al.[135] showed that benzene oxide, the presumed initial metabolite of benzene, is mutagenic for *S. typhimurium*.

There is considerable evidence that benzene can induce chromosome damage. Pollini and Colombi[136,137] first reported benzene-induced chromosome damage based on results from cultured bone marrow and lymphocytes from patients displaying serious benzene-induced bone marrow depression. The cultures displayed a high rate of aneuploid cell production. Both numerical and structural chromosome changes have been reported in lymphocytes cultured from the blood of workers whose exposure to benzene varied widely with respect to time of exposure and air levels of benzene.[138,139] Forni and co-workers[140-142] reported both stable and unstable chromosome aberrations. During follow-up studies, the unstable types seemed to disappear whereas the stable type persisted. Extensive chromosome abnormalities and abnormal clone formation were observed in patients displaying benzene-associated leukemia.[142]

Kahn and Kahn[143] and Picciano,[144] who investigated the cytogenic effects of prior industrial exposure to benzene, found increased numbers of chromosomal breaks and marker chromosomes including rings, dicentrics, translocations, and exchange figures in peripheral blood cells. Sarto and co-workers[145] examined 22 healthy workers from a benzene, toluene, and

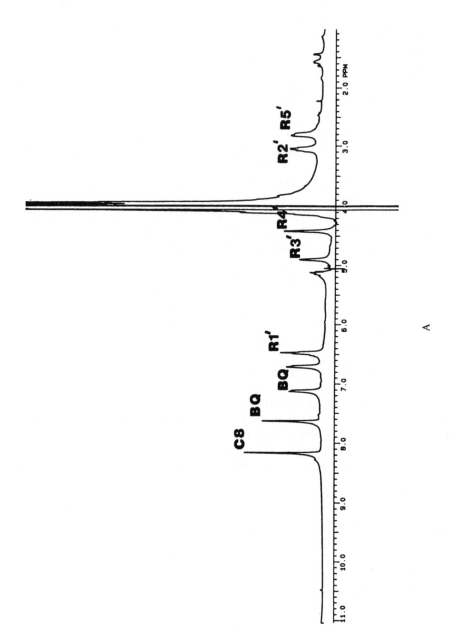

FIGURE 4.   NMR spectrum of deoxyguanosine adduct (A) and mass spectral analysis (B) of the major DNA adduct formed from in vitro incubation of hydroquinone and calf thymus DNA with iron added as a mild oxidizing agent. For the NMR spectrum, dioxone was the internal reference, number of transmissions = 500, and T = 22°C. The adduct was derivatized with TMS and analyzed on a Finnigan MAT mass spectrometer.

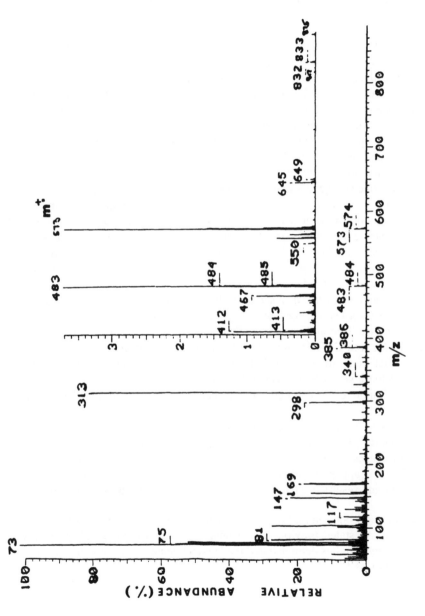

FIGURE 4B.

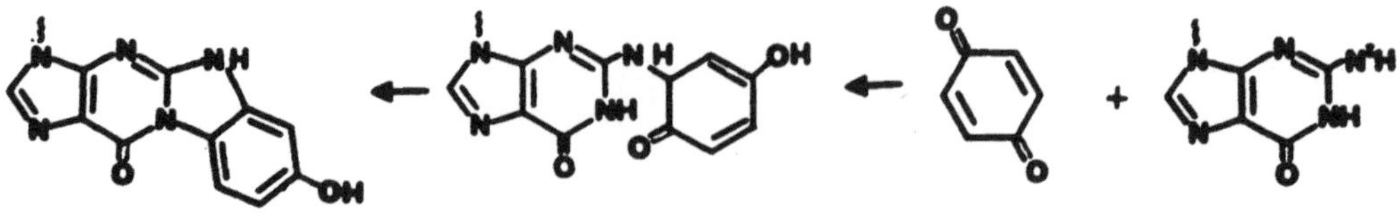

FIGURE 5. Proposed structure of hydroquinone derived deoxyguanosine adduct formed in vitro following addition of iron as a mild oxidizing agent.

xylene production factory. They reported no significant increase in SCE in the benzene group, but did report significantly higher chromosome-type aberrations.

In this area of investigation, unlike direct studies of production of leukemia by benzene, animal studies have been successful. Kissling and Speck[146,147] have produced chromosome aberrations in rabbits made pancytopenic with benzene. Pollini and Biscobaldi[148] have demonstrated mitotic arrest and unusual chromosomes and nuclei in cells under the influence of benzene. Inhibition of uptake of tritiated thymidine into DNA in bone marrow has been shown by Moeschlin and Speck[107] and Boje et al.[149] The benzene exposure was insufficient to produce chromosomal aberrations in the animals.

Tice et al.[150,151] demonstrated excessive sister chromatid exchanges (SCE) in DBA/2 mice exposed to 3100 ppm benzene by inhalation for 4 hr. The SCE frequency, which was approximately doubled, occurred at a level of exposure that did not cause chromosomal aberrations. By comparing the number of SCE found in successive generations of cultured marrow cells, it was determined that SCE occurred both as a result of exposure of previously unexposed cells to benzene and because of the persistence of lesions from a previous generation of cells. Tice and co-workers also observed an inhibition of bone marrow cellular proliferation.[150] The latter is a critical observation for the leukemogenic action of benzene because persistence of such lesions correlates with carcinogenesis in other systems.[152,153]

In a later set of studies Tice and co-workers[150] demonstrated that a 4-hr exposure to 28 ppm benzene in DBA/z and C57B1/6 mice resulted in a significant increase in SCE. They also demonstrated that pretreatment with inducers, partial hepatectomy, and competitive substrates decreased the level of SCE, which demonstrated a requirement for metabolism. These results were similar to those found for both benzene metabolism and ^{59}Fe reported by Sammet et al.[28]

Erexson et al.[154] exposed human peripheral blood lymphocytes to benzene, phenol, catechol, 1,2,4-benzenetriol, hydroquinone, and benzoquinone and showed a dose-dependent increase in SCE, decreases in mitotic indices, and inhibition of cell cycle kinetics.

From these studies it is apparent that benzene, although not a mutagen in the Ames assay, is a clastogenic agent and results in SCE. This observation has been observed in animals as well as humans.

Detailed information on both animal and human carcinogenicity is provided in Chapters 3 and 6, respectively. As stated previously, benzene is a known human leukemogen and only recently have animal studies demonstrated the leukemogenic nature of benzene and the occurrence of solid tumors.[155-157]

IV. CONCLUSION

Benzene exposure results in two distinct disease states (aplastic anemia and leukemia) that may be linked or may be two separate events resulting from different target sites and possibly different benzene metabolites. Aplastic anemia can be readily reproduced in animal models, but even in this instance, a definitive mechanism has not been found. The polyhydroxylated benzenes have been shown to interfere with a number of cellular functions and specifically damage organelles which could result in bone marrow hypocellularity.

Currently, there is no animal model for the leukemogenic action of benzene, although recent animal carcinogenicity studies do show some promise.

It has become apparent in recent years that the production of both toxic and carcinogenic effects of chemicals in many instances is due to the production of highly reactive metabolites which covalently bind to protein and nucleic acids, thereby altering cell function.[129,158] The data strongly indicate that one or more metabolites of benzene mediate benzene toxicity.[25] Inhibition of benzene metabolism by competitive inhibitors or partial hepatectomy protect against benzene toxicity.[13,28] Furthermore, benzene metabolites inhibit protein and RNA synthesis by covalently binding to cellular DNA.[63,64,128]

A comparison of the results of Post and Snyder,[58,59] Gilmour and Snyder,[52,53] and Sawahata and Neal,[74] using microsomal systems, suggest a similarity of the enzyme(s) responsible for the metabolism of benzene and phenol. Griffiths et al.[49,50] and Snyder[2] using isolated, reconstituted phenobarbital-induced rat hepatic mixed function oxidase suggest that the first step in benzene metabolism is predominantly mediated by cytochrome P-450 with the formation of an epoxide which rearranges to form phenol. Further metabolism is largely via free radical mechanisms. Treatment both in vivo and in vitro with free radical scavengers have demonstrated that the amount of covalent binding and adverse effects in vivo are prevented.

In addition to cytochrome P-450, other enzymes and nonenzymatic systems have been implicated in the further metabolism of phenol: myloperoxidase, DT-diapherase, oxidative burst, and autoxidation.

The ultimate target for benzene has yet to be described. In a general sense, the metabolite(s) which mediates toxicity/carcinogenesis must have an effect on the bone marrow. The most dramatic effects, from a mechanistic point of view, have been observed by Kalf et al.[63] and Rushmore et al.[64] who demonstrated that the covalent binding of benzene metabolites to DNA of mitochondria inhibited RNA synthesis. A similar inhibition of RNA synthesis by benzene metabolites was observed in nuclei by Post et al.[128] who suggested a similar mechanism. Jowa[159] has reported on specific DNA adducts to guanine and to other bases. In the case of benzene, the cytogenetic studies are quite compelling. Benzene has been consistently shown to cause chromosome breaks and an increase in SCE and red cell micronuclei in a series of studies over many years. Most of the recent reports have been devoted to SCE and micronucleus production. Although benzene is effective in producing these changes, it is not clear what the relationship, if any, is between these effects and leukemia at low doses of benzene.

The process by which benzene induces carcinogenesis appears to be extremely complicated. Despite numerous attempts at animal bioassays, the production of benzene-induced leukemia in animals has not been readily achieved. The studies of Snyder and co-workers[88,130] and Maltoni and co-workers[156] demonstrate that in the first instance extremely few leukemias can be induced and in the second instance other types of unrelated tumors were induced. It was not until the recent work of Cronkite et al.[155] that a high percentage of a leukemia variant was produced in animals following benzene treatment. This was accomplished by treating the animals for a short period of time and then allowing them time to recover. It is possible that in previous benzene carcinogenicity studies, the long period of time during which benzene was administered not only produced leukemia, but benzene also exerted cell replication inhibitory activity, counteracting leukemogenesis. This may be a general property of benzene carcinogenesis.

Cytogenetic and mutagenic studies, regardless of the dose of benzene employed, are indicators of mutagenic and possible carcinogenic potential only. In the case of benzene, the cytogenetic studies are quite compelling. Benzene has been consistently shown to cause chromosome breaks and an increase in SCE and red cell micronuclei in a series of studies over many years. Most of the recent reports have been devoted to SCE and micronucleus production. Although benzene is effective in producing these changes, it is not clear what the relationship, if any, is between these effects and leukemia at low doses of benzene. For example, Erexson et al.[154] indicated that catechol was the most potent metabolite for the formation of SCE whereas Post et al.[128] found that for the inhibition of MRNA synthesis in bone marrow nuclei, hydroquinone and benzoquinone were more potent than catechol. There is no evidence on the metabolite most likely to cause leukemia, but the major DNA adducts resulting from benzene are the hydroquinone and benzoquinone adducts on guanine.[159] Thus, the production by various metabolites of benzene of different end points, as proposed by Tice et al.,[150] suggests that there is not necessarily a sequential series of events leading from one or more forms of clastogenic changes to the production of leukemia. If clastogenic changes are parallel but independent of leukemogenesis, the dose effect relationships would be different, and the pharmacokinetics would be different.

If the theory that cancer is caused by chemically- or physically-induced autosomal mutations in DNA (somatic mutation theory) is correct, benzene should be mutagenic. However, benzene has not been shown to be mutagenic in short-term tests of mutagenicity.[134] If a mutagen is a chemical that causes a change in the DNA of a cell which leads to a phenotypic change in the cell, benzene might be termed a ''potential'' mutagen because it binds to DNA. However, although the binding is known to reduce RNA and protein synthesis, it has not been linked to the transformation of the affected cell to a cancer cell. Thus, we have yet to work out the sequence of events between benzene-adduct formation and the eventual development of neoplastic disease. It can be said that in given test systems, benzene can be shown to be clastogenic, but not mutagenic, and it can be shown to induce some types of cancer. However, there is as of yet no direct link between clastogenic effects and carcinogenic effects of benzene.

An underlying question which has yet to be adequately dealt with is: must frank bone marrow damage precede a leukemogenic response to benzene? It is clear that a terminal event following aplastic anemia in man is often some form of leukemia. If aplastic anemia or other severe bone marrow damage is a precursor to benzene-induced leukemia, it would appear that reasonably high exposures to benzene, i.e., doses high enough to produce severe bone marrow depression, would be necessary to induce leukemia, even though some clastogenic responses could be produced at lower doses. It is agreed that the production of benzene toxicity in man occurs at high levels of exposure; it has been readily accepted by all parties that continued exposure at or above 50 ppm for prolonged periods of time leads to bone marrow depression.[160] It is also now clear that hematological responses to benzene (suppressed stem cell activity, decreased LPs-induced B cell mitogenesis, and increases in SCE) can be elicited in animal test systems at atmospheric benzene concentrations of 10 ppm.[110,111,154] These observations, while not examples of aplastic anemia or leukemia, can be interpreted to indicate one of two eventualities. In the first, they may indicate the early stages of disease processes which will eventually lead to either of these diseases if exposure continues over a prolonged period of time. Another equally valid interpretation is that these are reversible biological responses which need not necessarily lead to aplastic anemia or leukemia because the dose is too low and normal repair mechanisms would not permit progression of the disease processes. We feel that the toxic effects observed both in animals and man are due to a metabolite or metabolites of benzene interacting at multiple target sites, and the end results are expressed as either aplastic anemia and/or leukemia.

REFERENCES

1. **Snyder, R. and Kocsis, J. J.**, Current concepts of chronic benzene toxicity, *Crit. Rev. Toxicol.*, 3, 256, 1975.
2. **Snyder, R.**, The benzene problem in historical perspective, *Fundam. Appl. Toxicol.*, 4, 692, 1983.
3. **Aksoy, M. and Erdem, S.**, Follow-up study on the mortality and the development of leukemia in 44 pancytopenia patients with chronic exposure to benzene, *Blood*, 52, 285, 1978.
4. **Aksoy, M., Erdem, S., and Dincol, G.**, Types of leukemia in chronic benzene poisoning. A study in thirty-four patients, *Acta Haematol.*, 55, 65, 1976.
5. **Maltoni, C.**, Myths and facts in the history of benzene carcinogenicity, in *Carcinogenicity and Toxicity of Benzene*, Mehlman, M. A., Ed., Princeton Scientific Publishers, Princeton, N.J., 1986, chap. 1.
6. **Mara, S. J. and Lee, S. S.**, Human exposure to atmospheric benzene, Stanford Res. Inst. Cent. for Resource and Environmental Systems Studies Rep., No. 30, October 1977, p. 151.
7. **Lee, S. D., Dourson, Mukerjee, D., Stara, J. F., and Kawecki, J.**, Assessment of benzene health effects in ambient water, in *Carcinogenicity and Toxicity of Benzene*, Mehlman, M. A., Ed., Princeton Scientific Publishers, Princeton, N.J., 1986, chap. 8.
8. **Day, E. A. and Anderson, D. F.**, Gas chromatographic and mass spectral identification of natural components of the aroma fraction of blue cheese, *J. Agric. Food Chem.*, 13, 2, 1965.
9. **Wilkens, W. F. and Lin, F. M.**, Gas chromatographic and mass spectral analyses of soybean milk volatiles, *J. Agric. Food Chem.*, 18, 333, 1970.
10. **Cleland, J. G. and Kingsbury, G. L.**, Multimedia environmental gasoline environmental assessment, EPA 60017-77-136, U.S. Environmental Protection Agency, Washington, D.C., 1977.
11. **Schrenk, H. H.**, Absorption, distribution and elimination of benzene from body tissues and fluids of dogs exposed to benzene vapor, *J. Int. Toxicol.*, 23, 20, 1941.
12. **Parke, D. V. and Williams, R. T.**, Studies in detoxication. The metabolism of benzene containing ^{14}C benzene, *Biochem. J.*, 54, 231, 1953.
13. **Andrews, L. S., Lee, E. W., Witmer, C. M., Kocsis, J. J., and Snyder, R.**, Effects of toluene on the metabolism, disposition and hemopoietic toxicity of ^{3}H benzene, *Biochem. Pharmacol.*, 26, 293, 1977.
14. **Rickert, D. E., Baker, T. S., Bus, J. S., Barrow, C. S., and Irons, R. D.**, Benzene disposition in the rat after exposure by inhalation, *Toxicol. Appl. Pharmacol.*, 49, 417, 1979.
15. **Sato, A., Fujiwara, Y., and Hirosawa, K.**, Pharmacokinetics of benzene and toluene, *Int. Arch. Arbeitsmed.*, 33, 169, 1974.
16. **Sato, A., Fujiwara, Y., and Nakajima, T.**, Solubility of benzene, toluene and M-xylene in various body fluids and tissues of rabbits, *Jpn. J. Ind. Health*, 16, 30, 1974.
17. **Sato, A. and Nakajima, T.**, A vial-equilibrium method to evaluate the drug-metabolizing enzyme activity for volatile hydrocarbons, *Toxicol. Appl. Pharmacol.*, 47, 41, 1979.
18. **Timbrell, J. A. and Mitchell, J. R.**, Toxicity-related changes in benzene metabolism *in vivo*, *Xenobiotica*, 7, 415, 1977.
19. **Hunter, O.**, *The Diseases of Occupations*, Little, Brown, Boston, 1962, 494.
20. **Sato, A.**, Toxicokinetics of benzene, toluene and xylenes, in *Environmental Carcinogens Selected Methods of Analysis*, Vol. 10, International Agency for Research on Cancer, Lyon, France, chap. 3, in press.
21. **Lazarew, N. W., Brussellowskaja, J. N., Laurow, F. B., and Lifschitz, F. B.**, Cutaneous permeability for petroleum ether and benzene, *Arch. Hyg.*, 106, 112, 1931.
22. **Cesaro, A. W.**, Is percutaneous absorption of benzene possible, *Med. Lav.*, 4, 151, 1946.
23. **Conca, G. C. and Mallagliate, A.**, Study of the percutaneous absorption of benzene, *Med. Lav.*, 46, 194, 1955.
24. **Sherwood, R. J.**, Benzene: the interpretation of monitoring results, *Ann. Occup. Hyg.*, 15, 409, 1972.
25. **Sherwood, R. J. and Carter, F. W. G.**, The measurement of occupational exposure to benzene vapor, *Ann. Occup. Hyg.*, 13, 125, 1970.
26. **Nomiyama, K. and Nomiyama, H.**, Respiratory elimination of organic solvents in man, *Int. Arch. Arbeitsmed.*, 32, 85, 1974.
27. **Nomiyama, K. and Nomiyama, H.**, Respiratory retention, uptake and excretion of organic solvents in man, *Int. Arch. Arbeitsmed.*, 32, 75, 1974.
28. **Sammet, D., Lee, E. W., Kocsis, J. J., and Snyder, R.**, Partial hepatectomy reduces both metabolism and toxicity of benzene, *J. Toxicol. Environ. Health*, 5, 785, 1979.
29. **Greenlee, W. F., Sun, J. D., and Bus, J. S.**, A proposed mechanism of benzene toxicity: formation of reactive intermediates from polyphenol metabolites, *Toxicol. Appl. Pharmacol.*, 59, 187, 1981.
30. **Schultzen, O. and Naunyn, B.**, Wer das Verbalten des Kohlenwasserstoffes im Organisms, *Arch. Anat. Physiol.*, 1, 349, 1867.
31. **Nencki, M. and Giacosa, P.**, Uber die Oxydation die Aromatischen Wasserstaffe in Turkoyser, *Z. Physiol. Chem.*, 4, 325, 1880.

32. **Porteous, J. W. and Williams, R. T.**, Studies in detoxication. The metabolism of benzene. The isolation of phenol, catechol, quinol and hydroxyquinol from ethereal sulfate fraction of the urine of rabbits receiving benzene orally, *Biochem. J.*, 44, 56, 1949.

33. **Porteous, J. W. and Williams, R. T.**, Studies in detoxication. The metabolism of benzene. The determination of phenol in urine with 216-dichloroquenone chloroimide, *Biochem. J.*, 44, 46, 1949.

34. **Abouel-Marakem, M. M., Milburn, P., Smith, R. C., and Williams, R. T.**, Biliary excretion of foreign compounds: benzene and its derivatives in the rat, *Biochem. J.*, 105, 1269, 1967.

35. **Cornish, H. N. and Ryan, R. C.**, Metabolism of benzene in non-fasted, fasted and arylhydrocarbon inhibited rats, *Toxicol. Appl. Pharmacol.*, 7, 767, 1965.

36. **Snyder, R., Lee, E. W., Kocsis, J. J., and Witmer, C. M.**, Bone marrow depressant leukomogenic action of benzene, *Life Sci.*, 21, 1709, 1977.

36a. **Bolczak, L. and Nerland, D. E.**, Inhibition of erythropoiesis by benzene and benzene metabolites, *Toxicol. Appl. Pharmacol.*, 69, 363, 1983.

37. **Witz, G. W., Rao, G. S., and Goldstein, B. D.**, Short-term toxicity of trans, trans-muconaldehyde, *Toxicol. Appl. Pharmacol.*, 80, 511, 1985.

38. **Latriano, L., Goldstein, B. D., and Witz, G.**, Formation of muconaldehyde from ^{14}C-benzene in mouse liver microsomes, *Toxicologist*, 6, 153, 1986.

39. **Yant, W. P., Schrenk, H. H., Sayers, R. R., Howarth, A. A., and Rwinhert, W. H.**, Urine sulfate determinations as a measure of benzene exposure, *J. Ind. Hyg. Toxicol.*, 18, 69, 1936.

40. **Hammond, J. W. and Herman, E. R.**, Industrial hygiene features of a petro-chemical benzene plant design and operation, *Am. Ind. Hyg. Assoc. J.*, 21, 173, 1960.

41. **Teisenger, J. and Skramovsky, S.**, The metabolism of benzene in man, *Procovni Lek.*, 4, 175, 1952.

42. **Bolcsak, L. E. and Nerland, D. E.**, Inhibition of erythropoiesis by benzene and benzene metabolites, *Toxicol. Appl. Pharmacol.*, 69, 363, 1983.

43. **Nomiyama, K.**, Studies on the poisoning by benzene and its homologues oxidation rate of benzene and benzene poisoning, *Med. J. Shinshu Univ.*, 7, 41, 1962.

44. **Tunek, A., Platt, K. L., Bentley, P., and Oesch, F.**, Microsomal metabolism of benzene to species irreversibly binding to microsomal protein and effects of modification of this metabolism, *Mol. Pharmacol.*, 14, 920, 1978.

45. **Daly, J. W., Jerina, D. M., and Witkop, B.**, Arene oxides and the NIH shift: the metabolism, toxicity and carcinogenicity of aromatic compounds, *Experimentia*, 28, 1129, 1972.

46. **Ingelman-Sundberg, M. and Hagbjork, A. L.**, On the significance of the cytochrome P-450-dependent hydroxyl radical-mediated oxygenation mechanism, *Xenobiotica*, 12, 673, 1982.

47. **Johansson, I. and Ingelman-Sundberg, M.**, Hydroxyl radical-mediated, cytochrome P-450-dependent metabolic activation of benzene in microsomes and reconstituted enzyme systems from rabbit liver, *J. Biol. Chem.*, 258, 7311, 1983.

48. **Cavalieri, E. L., Rogan, E. G., Roth, R. W., Saugier, R. K., and Hakam, A.**, The relationship between ionization potential and horseradish peroxidase/hydrogen peroxide catalyzed binding of aromatic hydrocarbons to DNA, *Chem. Biol. Interact.*, 47, 87, 1983.

49. **Griffiths, J. C., Kalf, G. F., and Snyder, R.**, The metabolism of benzene by a reconstituted purified phenobarbital-induced rat liver mixed function oxidase, *Fed. Proc. Fed. Am. Soc. Exp. Biol.*, 43, 934, 1984.

50. **Griffiths, J. C., Kalf, G. F., and Snyder, R.**, The metabolism of phenol by a reconstituted purified phenobarbital-induced rat liver mixed function oxidase system, *Fed. Proc. Fed. Am. Soc. Exp. Biol.*, 44, 516, 1985.

51. **Sawahata, T. and Neal, R. A.**, Biotransformation of phenol to hydroquinone and catechol by rat liver microsomes, *Mol. Pharmacol.*, 23, 453, 1983.

52. **Gilmour, S. and Snyder, R.**, Metabolism of benzene and phenol in rat liver microsomes, *Fed. Proc. Fed. Am. Soc. Exp. Biol.*, 42, 1136, 1983.

53. **Gilmour, S. and Snyder, R.**, Similarities in the microsomal metabolism of benzene and its metabolite phenol, *Pharmacologist*, 25, 210, 1983.

54. **Tunek, A., Platt, K. L., Przyblyski, M., and Oesch, F.**, Multi-step metabolic activation of benzene. Effect of superoxide dismutase on covalent binding to microsomal macromolecules, and identification of glutathione conjugates using high pressure liquid chromatography and field desorption mass spectrometry, *Chem. Biol. Interact.*, 33, 1, 1980.

55. **Lunte, S. M. and Kissinger, P. T.**, Detection and identification of sulfhydryl conjugates of p-benzoquinone in microsomal incubations of benzene and phenol, *Chem. Biol. Interact.*, 47, 195, 1983.

56. **Wallin, H., Melin, P., Schelin, C., and Jergel, B.**, Evidence that covalent binding of metabolically activated phenol to microsomal proteins is caused by oxidized products of hydroquinone and catechol, *Chem. Biol. Interact.*, 55, 335, 1985.

57. **Gonasun, L. M., Witmer, C. M., Kocsis, J. J., and Snyder, R.**, Benzene metabolism in mouse liver microsomes, *Toxicol. Appl. Pharmacol.*, 26, 398, 1973.

58. **Post, G. B., and Snyder, R.,** Fluoride stimulation of microsomal benzene metabolism, *Toxicol. Environ. Health,* 11, 799, 1983.

59. **Post, G. B. and Snyder, R.,** Effects of enzyme induction on microsomal benzene metabolism, *J. Toxicol. Environ. Health,* 11, 811, 1983.

60. **Andrews, L. S., Sasame, H. A., and Gillette, J. R.,** ^{3}H Benzene metabolism in rabbit bone marrow, *Life Sci.,* 25, 567, 1979.

61. **Gollmer, L., Graf, H., and Ullrich, V.,** Characterization of benzene monooxygenase system in rabbit bone marrow, *Biochem. Pharmacol.,* 33, 3597, 1984.

62. **Irons, R. D., Heck, H. D., Moore, B. J., and Muirhead, K. A.,** Effects of short-term benzene administration on bone marrow cell cycle kinetics in the rat, *Toxicol. Appl. Pharmacol.,* 51, 399, 1979.

63. **Kalf, G. F., Rushmore, T., and Snyder, R.,** Benzene inhibits RNA and protein synthesis in mitochondria from regenerating liver and bone marrow, *Chem. Biol. Interact.,* 42, 371, 1982.

64. **Rushmore, T., Snyder, R., and Kalf, G.,** Covalent binding of benzene and its metabolites to DNA in rabbit bone marrow mitochondria *in vitro, Chem. Biol. Interact.,* 49, 133, 1984.

65. **Deisseroth, A. and Dounce, A. L.,** Catalase: physical and chemical properties, mechanism of catalysis and physiological role, *Physiol. Rev.,* 50, 319, 1970.

66. **Chance, B. and Maehly, A. C.,** Assay of catalases and peroxidases, in *Methods in Enzymology,* Vol. 11, Colowick, S. P. and Kaplan, N. O., Eds., Academic Press, New York, 1955, 764.

67. **Agner, K.,** Verdoperoxidase, *Acta Physiol. Scand.,* 2 (Suppl. 8), 1941.

68. **Agner, K.,** Crystalline myeloperoxidase, *Acta Chem. Scand.,* 12, 89, 1958.

69. **Kimura, S., Elce, J. S., and Jellinck, P. H.,** Immunological relationship between peroxidases in eosinophils, uterus and other tissues of the rat, *Biochem. J.,* 213, 165, 1983.

70. **Mandel, I. D. and Ellison, S. A.,** The biological significance of nonimmunoglobulin defense factors, *The Lactoperoxidase System: Chemistry and Biological Significance,* Pruitt, K. M. and Tenovuo, J. O., Eds., Marcel Dekker, New York, 1985, 1.

71. **Olsen, R. L. and Little, C.,** Purification and some properties of myeloperoxidase and eosinophil peroxidase from human blood, *Biochem. J.,* 209, 781, 1983.

72. **Andrews, P. C., Parnes, C., and Krinsky, N. I.,** Comparison of myeloperoxidase and hemi-myeloperoxidase with respect to catalysis, regulation and bacterial activity, *Arch. Biochem. Biophys.,* 228, 439, 1984.

73. **Pember, S. O., Shapira, R., and Kinkade, J. M., Jr.,** Multiple forms of myeloperoxidase from human neutrophilic granulocytes: evidence for differences in compartmentalization, enzymatic activity, and subunit structure, *Arch. Biochem. Biophys.,* 221, 391, 1983.

74. **Sawahata, T. and Neal, R. A.,** Horseradish peroxidase-mediated oxidation of phenol, *Biochem. Biophys. Res. Commun.,* 109, 988, 1982.

75. **Smart, R. C. and Zannoni, V. G.,** DT-diaphorase and peroxidase influence the covalent binding of the metabolites of phenol, the major metabolite of benzene, *Mol. Pharmacol.,* 26, 105, 1984.

76. **Smart, R. C. and Zannoni, V. G.,** Effect of ascorbate on covalent binding of benzene and phenol metabolites to isolated tissue preparations, *Toxicol. Appl. Pharmacol.,* 77, 334, 1985.

77. **G'Brien, P. J.,** One electron oxidation in cell toxicity, in *Microsomes and Drug Oxidations: Proc. 6th Int. Symp.,* James R. Gillette, Boobis, A. R., Caldwell, J., De Matteis, F., and Elcombe, C. R., Eds., Taylor and Francis, London, 1985, 284.

78. **Soberman, R. J. and Karnovsky, M. L.,** Biochemical properties of macrophages, *Lymphokines. A Forum for Immunoregulatory Cell Products,* Pick, E., Ed., Academic Press, New York, 1981.

79. **Adams, D. O. and Hamilton, T. A.,** The cell biology of macrophage activation, *Annu. Rev. Immunol.,* 2, 283, 1984.

80. **McPhail, L. C. and Snyderman, R.,** Mechanisms of regulating the respiratory burst, *Regulation of Leukocyte Function,* Snyderman, R., Ed., Plenum Press, New York, 1984, 247.

81. **Babior, B. M.,** Oxidants from phagocytes: agents of defense and destruction, *Blood,* 64, 959, 1984.

82. **Babior, B. M.,** Superoxide and oxidative killing by phagocytes, *The Biology and Chemistry of Active Oxygen,* Bannister, J. V. and Bannister, W. H., Eds., Elsevier, Amsterdam, 1984, 190.

83. **Kappus, H. and Sies, H.,** Toxic drug effects associaed with oxygen metabolism: redox cycling and lipid peroxidation, *Experientia,* 37, 1233, 1981.

84. **Lorentzen, R. J., Lesko, S. A., McDonald, K., and Ts'O, P. O.,** Toxicity of metabolic benzo(a)pyrenediones to cultured cells and the dependence upon molecular oxygen, *Cancer Res.,* 39, 3194, 1979.

85. **Lorentzen, R.,** Benzo(a)pyrenedione/benzo(a)pyrenehydroquinone redox couples and their biological significance, *Proc. Am. Assoc. Cancer Res.,* 17, 100, 1976.

86. **Lorentzen, R., Caspary, W., Lesko, S., and Ts'O, P.,** The autoxidation of 6-hydroxybenzo(a)pyrene and 6-oxobenzo(a)pyrene radical, reactive metabolites of benzo(a)pyrene, *Biochemistry,* 14, 3970, 1975.

87. **Deichmann, W. B., MacDonald, W. E., and Bernal, E.,** Hemopoeic tissue toxicity of benzene vapors, *Toxicol. Appl. Pharmacol.,* 5, 201, 1963.

88. **Snyder, C., Goldstein, B., Sellakumar, A., Bromberg, I., Laskin, S., and Albert, R.,** The inhalation toxicology of benzene: incidence of hematopoietic neoplasms and hematotoxicity in AKR/J and C57B1/6J mice, *Toxicol. Appl. Pharmacol.,* 54, 323, 1980.

89. **Goldstein, B. D.,** Clinical hematotoxicity of benzene, in *Carcinogenicity and Toxicity of Benzene,* Vol. 4, Mehlman, M. A., Ed., Princeton Scientific Publishers, Princeton, N.J., 1985, chap. 5.

90. **Selling, H.,** Benzo as a leucotoxin. Studies on the degeneration and regeneration of the blood and haematopoietic organs, *Johns Hopkins Hosp. Rep.,* 17, 83, 1916.

91. **Harigaya, K., Miller, M. E., Cronkite, E. P., and Drew, R. T.,** The detection of *in vivo* hematotoxicity of benzene by *in vitro* liquid bone marrow cultures, *Toxicol. Appl. Pharmacol.,* 60, 346, 1981.

92. **Green, J. P., Snyder, C. A., LaDue, J., Goldstein, B. D., and Albert, R. E.,** Acute and chronic dose/response effects of inhaled benzene on multipotential hematopoietic stem (CFU-S) and granulocyte/macrophage progenitor (GM-CFU-C) cells in CD-1 mice, *Toxicol. Appl. Pharmacol.,* 58, 492, 1981.

93. **Tunek, A., Hogstedt, B., and Olofsson, T.,** Mechanism of benzene toxicity. Effects of benzene and benzene metabolites on bone marrow cellularity, number of granulopoietic stem cells and frequency of micronuclei in mice, *Chem. Biol. Interact.,* 39, 129, 1982.

94. **Steinberg, B.,** Bone marrow regeneration in experimental benzene intoxication, *Blood,* 4, 550, 1949.

95. **Moeschlin, S. and Speck, B.,** Experimental studies on the mechanism of action of benzene on the bone marrow (radioautographic studies using ^{3}H thymidine), *Acta Haematol.,* 38, 104, 1967.

96. **Lee, E. W., Kocsis, J. J., and Snyder, R.,** The use of ferrokinetics in the study of experimental anemia, *Environ. Health Perspect.,* 39, 29, 1981.

97. **Frash, V. N., Yushkov, B. G., Karaulov, A. V., and Skuratov, V. L.,** Mechanism of action of benzene on hematopoiesis (investigation of hematopoietic stem cells), *Bull. Exper. Biol. Med.,* 82, 985, 1976.

98. **Irons, R. D., Neptun, D. A., and Pfeifer, R. W.,** Effects of the principal hydroxymetabolites of benzene: evidence for a common mechanism, *J. Reticuloendothelial Soc.,* 30, 359, 1981.

99. **Irons, R. D., Pfeifer, R. W., Aune, T. M., and Pierce, C. W.,** Soluble immune response suppressor (SIRS) inhibits microtubule function in vivo and microtubule assemble *in vitro, J. Immunol.,* 133, 2032, 1984.

100. **Cronkite, E., Inoue, T., Carsten, A., Miller, M., Bulles, J., and Drew, R.,** Effects of benzene inhalation on murine pluripotent stem cells, *J. Toxicol. Env. Health,* 9, 411, 1982.

101. **Pollini, G., Biscaldi, G. P., and Corsico, R.,** L'Azione del benzolo sull'attrita proliferativa delle callule emopoietiche embrionarie, *Med. Lav.,* 56, 738, 1965.

102. **D'Souza, M., Snyder, R., and Kocsis, J. J.,** Benzene inhibits ovarian hypertrophy in the hemispayed rat, Abstract A40, Soc. Toxicology 18th Annu. Meet., 1979.

103. **Lee, E. W., Kocsis, J. J., and Snyder, R.,** Dose dependent inhibition of ^{59}Fe incorporation into circulating erythrocytes, *Res. Commun. Chem. Pathol. Pharmacol.,* 5, 547, 1973.

104. **Lee, E. W., Kocsis, J. J., and Snyder, R.,** Acute effects of benzene on ^{59}Fe incorporation into circulating erythrocytes, *Toxicol. Appl. Pharmacol.,* 27, 431, 1974.

105. **Allen, T. D. and Dexter, T. M.,** The essential cells of the hemopoietic microenvironment, *Exp. Hematol.,* 12, 517, 1984.

106. **Uyeki, E. M., Ashkar, A. E., Shoeman, D. W., and Bisel, T. U.,** Acute toxicity of benzene inhalation to hematopoietic precursor cells, *Toxicol. Appl. Pharmacol.,* 40, 49, 1977.

107. **Moeschlin, S. and Speck, B.,** Experimental studies on the mechanism of action of benzene on the B.M., *Acta Haematol.,* 38, 104, 1967.

108. **Rondanelli, E. G., Gorini, P., Magliulo, E., Vannini, V., Fiori, G. P., and Parma, R.,** Benzene induced anomalies in mitotic cycle of living erythrocites, *Sangre,* 9, 342, 1964.

109. **Baarson, K., Snyder, C., Green, J., Sellakumar, A., Goldstein, B., and Albert, R.,** The hematotoxic effects of inhaled benzene on peripheral blood, bone marrow and spleen cells are increased by ingested ethanol, *Toxicol. Appl. Pharmacol.,* 64, 393, 1982.

110. **Baarson, K., Snyder, C., and Albert, R.,** Repeated exposure of C57B1 mice to inhaled benzene at 10 ppm markedly depressed erythropoietic colony formation, *Toxicol. Lett.,* 20, 337, 1984.

111. **Rozen, M. G., Snyder, C. A., and Albert, R. E.,** Depressions in B- and T-lymphocyte mitogen-induced blastogenesis in mice exposed to low concentrations of benzene, *Toxicol. Lett.,* 20, 343, 1984.

112. **Gram, T. E., Okine, L. K., and Gram, R. A.,** The metabolism of xenobiotics by certain extrahepatic organs and its relation to toxicity, *Ann. Rev. Pharmacol. Toxicol.,* 26, 259, 1986.

113. **Irons, R. D., Wierda, D., and Pfeifer, R. W.,** The immunotoxicity of benzene and its metabolites, in *Carcinogenicity and Toxicity of Benzene,* Vol. 4, Mehlman, M. A., Ed., Princeton Scientific Publishers, Princeton, N.J., 1985, chap. 4.

114. **Pfeifer, R. W. and Irons, R. D.,** Inhibition of lectin-stimulated lymphocyte agglutination and mitogenesis of hydroquinone reactivity with intracellular sulfhydryl groups, *Exp. Mol. Pathol.,* 35, 189, 1981.

115. **Hesketh, P. J., Sullivan, R., Valeri, C. R., and McCarroll, L. A.,** The production of granulocyte-monocyte colony-stimulating activity by isolated human T lymphocyte subpopulations, *Blood,* 63, 1141, 1984.

116. **Verma, D. S., Johnson, D. A., and McCredie, K. B.,** Identification of T lymphocyte subpopulations that regulate elaboration of granulocyte-macrophage colony stimulating factor, *Br. J. Haematol.*, 57, 505, 1984.

117. **Prystowsky, M. B., Otten, G., Naujokas, M. F., Vardiman, J., Ihle, J. N., Goldwasser, E., and Fitch, F. W.,** Multiple hemopoietic lineages are found after stimulation of mouse bone marrow precursor cells with interleukin 3, *Am. J. Pathol.*, 117, 171, 1984.

118. **Goldwasser, E., Ihle, J., Prystowsky, M., Rich, I., and Van Zant, G.,** *Normal and Neoplastic Hematopoiesis*, Alan R. Liss, New York, 1983, 129.

119. **Wierda, D., Irons, R. D., and Greenlee, W. F.,** Immunotoxicity in C57B1/6 mice exposed to benzene and aroclor 1254, *Toxicol. Appl. Pharmacol.*, 60, 410, 1981.

120. **Irons, R. D. and Moore, B. J.,** Effect of short term benzene administration on circulating lymphocyte subpopulations in the rabbit: evidence of selective B-lymphocyte sensitivity, *Res. Commun. Chem. Pharmacol.*, 27, 147, 1980.

121. **Greenlee, W. F. and Irons, R. D.,** Modulation of benzene-induced lymphocytopenia in the rat by 2,4,5,2′,4′,5′-hexachlorobiphenyl and 3,4,3′,4′-tetrachlorobiphenyl, *Chem. Biol. Interact.*, 33, 345, 1981.

122. **Irons, R. D. and Neptun, D. A.,** Effects of the principal hydroxy-metabolites of benzene on microtubule polymerization, *Arch. Toxicol.*, 45, 297, 1980.

123. **Garnett, H. M., Cronkite, E. P., and Drew, R. T.,** Effect of *in vivo* exposure to benzene on the characteristics of bone marrow adherent cells, *Leukemia Res.*, 7, 803, 1983.

124. **Gaido, K. W. and Wierda, D.,** In vitro effects of benzene metabolites on mouse bone marrow stromal cells, *Toxicol. Appl. Pharmacol.*, 76, 45, 1984.

125. **Haak, H. L. and Speck, B.,** Inhibition of CFU_E and BFU_E by mononuclear peripheral blood cells during chronic benzene treatment in rabbits, *Acta Haemat.*, 67, 27, 1982.

126. **Kalf, G. F., Rushmore, T., and Snyder, R.,** Benzene inhibits RNA synthesis in mitochondria from liver and bone marrow, *Chem. Biol. Interact.*, 42, 353, 1982.

127. **Schwartz, C. S., Snyder, R., and Kalf, G. F.,** The inhibition of mitochondrial DNA replication *in vitro* by the metabolites of benzene, hydroquinone and p-benzoquinone, *Chem. Biol. Interact.*, 53, 327, 1985.

128. **Post, G. B., Snyder, R., and Kalf, G. F.,** Inhibition of mRNA synthesis in rabbit bone marrow nuclei *in vitro* by quinone metabolites of benzene, *Chem. Biol. Interact.*, 50, 203, 1984.

129. **Jollow, D. J., Mitchell, J. R., Potter, W. Z., Davis, D. C., Gillette, J. R., and Brodie, B. B.,** Acetaminophen-induced hepatic necrosis. II. Role of covalent binding *in vivo*, *J. Pharmacol. Exp. Ther.*, 187, 195, 1973.

130. **Snyder, R., Parke, D. V., Kocsis, J. J., Jollow, D. J., Gibson, G. G., and Witmer, C. M., Eds.,** *Biological Reactive Intermediates II: Chemical Mechanisms and Biological Effects*, Plenum Press, New York, 1982.

131. **Lutz, W. K. and Schlatter, C.,** Mechanism of the carcinogenic action of benzene: irreversible binding to rat liver DNA, *Chem. Biol. Interact.*, 18, 241, 1977.

132. **Rushmore, T. H.,** The Metabolism of Benzene and its Inhibition of Macromolecular Synthesis in Mitochondria, Dissertation, Thomas Jefferson University, Philadelphia, Pa., 1983, 157.

133. **Nylander, P., Olaffson, H., Rasmuson, B., and Sahlin, H.,** Mutagenic effects of petrol in drosophila melanogastric, *Mutat. Res.*, 57, 163, 1978.

134. **Dean, B. J.,** Genetic toxicology of benzene, toluenes, xylenes and phenols, *Mutat. Res.*, 47, 75, 1978.

135. **Pulkrabek, P., Kinoshita, T., and Jeffery, A. M.,** Benzene oxide: in vitro mutagenic and toxic effects, *Proc. 16th Annu. Meet. Am. Soc. Chem. Oncol.*, 21, 107, 1980.

136. **Pollini, G. and Colombi, R.,** Il danna cromosomico medollare mell'hemopatia benenica and anemia aplastica benzolica, *Med. Lav.*, 55, 241, 1964.

137. **Pollini, G. and Colombia, R.,** Il danna cromosomico medollare mell'hemopatia benenica and anemia aplastica benzolico, *Med. Lav.*, 55, 641, 1964.

138. **Tough, I. and Brown, W. C.,** Chromosome aberrations and exposure to ambient benzene, *Lancet*, 1, 648, 1965.

139. **Tough, I., Smith, P., Brown, W. C., and Harnden, D.,** Chromosome studies on workers exposed to atmospheric benzene. The possible influence of age, *Eur. J. Cancer*, 6, 49, 1970.

140. **Forni, A., Capellini, A., Pacifico, E., and Vigliani, E.,** Chromosome changes and their evaluation in subjects with past exposure to benzene, *Arch. Environ. Health*, 23, 385, 1971.

141. **Forni, A., Pacifico, E., and Simonta, A.,** Chromosome studies in workers exposed to benzene or toluene or both, *Arch. Environ. Health*, 22, 373, 1971.

142. **Forni, A. and Moreo, L.,** Chromosome studies in a case of benzene-induced erythroleukemia, *Eur. J. Cancer*, 5, 459, 1969.

143. **Kahn, H. and Kahn, M. H.,** Cytogenetische Untugiechimgen bei Chromischen Benzol-Exposition, *Arch. Toxikol.*, 31, 39, 1973.

144. **Picciano, D.,** Cytogenetic study of workers exposed to benzene, *Environ. Res.*, 19, 33, 1979.

145. **Sarto, F., Cominato, I., Pinton, A., Brovedani, P., Merier, E., Peruzzi, M., Bianchi, V., and Lewis, A.,** Cytogenetic study on workers exposed to low concentrations of benzene, *Carcinogenesis,* 5, 827, 1984.

146. **Kissling, M. and Speck, B.,** Chromosome aberrations in experimental benzene intoxication, *Helv. Med. Acta,* 36, 59, 1971.

147. **Kissling, M. and Speck, B.,** Further studies on experimental benzene induced aplastic anemia, *Blut,* 25, 97, 1972.

148. **Pollini, G. and Biscobaldi, G. P.,** Cytogenetic effects following benzene exposure, *Med. Lav.,* 58, 205, 1967.

149. **Boje, V. H., Benkel, W., and Heiniger, H. J.,** Untersuchungen zur Blukopoese im Knockenmark der Rattenach Chonischer Benzol-Inhation, *Blut,* 21, 250, 1970.

150. **Tice, R. R., Vogt, T. F., and Costa, D. L.,** Cytogenetic effects of inhaled benzene on murine bone marrow, *Genotoxic Effects of Airborne Agents,* Tice, R. R., Costa, D. L., and Schaich, K. M., Eds., Plenum Press, New York, 1982, 257.

151. **Tice, R., Costa, D., and Drew, R.,** Cytogenetic effects of inhaled benzene in murine bone marrow: induction of sister chromatid exchanges, chromosomal aberrations, and cellular proliferation inhibition in DBA/2 mice, *Proc. Natl. Acad. Sci.,* 77, 2148, 1980.

152. **Goth, R. and Rajewsky, M. F.,** Persistence of O^6-ethyl guanine in rat brain DNA; correlation with nervous system — specific carcinogenesis ethylnitrosourea, *Proc. Natl. Acad. Sci. U.S.A.,* 71, 639, 1974.

153. **Nicoli, J. W., Swann, P. F., and Pegg, A. W.,** Effect of dimethylnitrosamine on persistance of methylated guanines in rat liver and kidney DNA, *Nature,* 254, 261, 1975.

154. **Erexson, G. L., Wilmer, J. L., and Kligerman, A. D.,** Sister chromatid exchange induction in human lymphocytes exposed to benzene and its metabolites *in vitro, Cancer Res.,* 45, 2471, 1985.

155. **Cronkite, E., Bullis, J., Inoue, T., and Drew, R.,** Benzene inhalation produces leukemia in mice, *Toxicol. Appl. Pharmacol.,* 75, 358, 1984.

156. **Maltoni, D., Conti, B., and Cotti, G.,** Benzene: a multipotential carcinogen. Results of long-term bioassays performed at the Bologna Institute of Oncology, *Am. J. Ind. Med.,* 4, 589, 1983.

157. **Huff, J.,** Technical report on the Toxicology and Carcinogenesis Studies of Benzene (CAS No. 71-43-2) In F344/N Rats and B6C3F$_1$ Mice (Gavage Studies). National Toxicology Program, Research Triangle Park, N.C., 1984.

158. **Brookes, P.,** Role of covalent binding on carcinogenicity, *Biological Reactive Intermediates,* Jollow, D. J., Kocsis, J. J., Snyder, R., and Vainio, H., Eds., Plenum Press, New York, 1977, 470.

159. **Jowa, L. B.,** Formation of DNA Adducts from the Benzene Metabolites, Hydroquinone and Benzoquinone, Dissertation, Department of Pharmacology and Toxicology, Rutgers State University, 1986, 1.

160. International Agency for Research On Cancer, Monographs on the evaluation of carcinogenic risk of chemicals to man: some industrial chemicals and dyestuffs, International Agency for research on Cancer, Lyon, France, pp. 93—148, 395—398, 1982.

Chapter 5

BENZENE HEMATOTOXICITY

Muzaffer Aksoy

TABLE OF CONTENTS

I. ACUTE BENZENE TOXICITY

Benzene (C_6H_6) is the parent hydrocarbon of the aromatic group, the resonant cyclic compound consisting of only carbon and hydrogen.[1,2] In 1825, Faraday was able to isolate benzene from coal tar naptha which was used as a rubber solvent.[3] Kekule was the first to determine the arrangement of six carbon atoms and the closed ring of the benzene nucleus in 1866. The term "benzene" is used in England and in the U.S. while benzol is favored in Germany. In France, benzene is called "benzine". Benzene is produced mainly in petrochemical operations, but small quantities may be obtained as a by-product of operations in the steel industry. Benzene stems not only from industrial sources. All crude oils contain small amounts of benzene, and vegetation emits benzene.[2] It is also found in tap and unprocessed water and in animal sources. Tobacco smoke contains benzene in insignificant amounts.[4-6] The discovery of aniline dye gave reason for industry to develop the distillation of coal tar to obtain benzene, homologs, and phenols.[3] Benzene is used as a solvent for rubber, gum, dyes, resins, fats, and alkaloids, besides being used in the manufacture of drugs, dyes, and explosives. In the last decade, the development of the chemical industry, especially in the field of plastics, firmly established the necessity of using large quantities of benzene as the starting material for chemical synthesis. Benzene is often used to clean machinery as well as grease from hands, and has even been used as a domestic agent. Additional uses of benzene are still being discovered; the possible use of benzene to replace lead as the "antiknock" component in motor oils is an example. However, public exposure to benzene in petroleum (ranging between 1 to 5%) has caused increased concern regarding benzene as a significant environmental pollutant.[4-10] All data concerning benzene occurrence and uses are discussed extensively in Chapter 1.

A. History

The first case of acute benzene toxicity was reported in England in 1862.[11] Following sucking at a syphon which did not draw, a worker swallowed benzene and, following unconsciousness, he died. At necropsy, the brain, lungs, and liver were full of blood. The second case of acute benzene toxicity had been reported again in England in 1889.[12] A man erroneously swallowed liquid benzene and, following this, he was unconscious. In his breath the odor of benzene was prominent. He was hospitalized and saved. Santesson[13] in his paper on benzene toxicity mentioned two fatal cases of acute benzene toxicity, one female and the other male.

B. Occurrence

Acute benzene toxicity usually occurs following inhalation of concentrated benzene vapor, usually in poorly ventilated rooms. Exposure to massive concentrations of benzene in the range of 2.5% by volume in air is rapidly fatal. A concentration of 19,000 to 20,000 ppm (61,693 to 64,940 mg/m³) is considered fatal to human beings after an exposure of a duration of 5 to 10 min.[14,15] When 5000 ppm (16,350 mg/m³) benzene is inhaled for $^1/_2$ to 1 hr, acute benzene toxicity may develop. Similarly, 7000 to 7500 ppm (22,729 to 24,352.5 mg/m³) is highly dangerous to life after an exposure duration of 30 min. On the other hand, inhalation of 700 to 3000 ppm (2272 to 9741 mg/m³) of benzene for of a duration of $^1/_2$ to 1 hr can be tolerated by human beings, but can cause unconsciousness after this period. Furthermore, severe acute benzene toxicity will occur if 1500 ppm (4870.5 mg/m³) is inhaled for 1 hr. Exposure to 500 ppm (1623.5 mg/m³) for 1 hr will cause clinically determinable changes, and exposure to 50 to 100 ppm (162.35 to 324.7 mg/m³) for 300 hr will cause headache, lethargy, and general exhaustion. Exposure to 2.5 ppm (8.1 mg/m³) for 480 min will not lead to any symptom. Some of these data are summarized in Table 1.

According to Hunter,[3] 13 fatal cases of acute benzene toxicity were recorded in England

Table 1
EFFECTS OF BENZENE VAPOR ON HUMAN BEINGS

Air concentration (ppm)	Exposure (min)	Short term effects
19,000—20,000	5—10	Fatal
7,500	30	Dangerous to life
3,000	30	Endurable
1,500	60	Serious symptoms
500	60	Symptoms of illness
50—150	300	Headache, lassitude, weariness
25	480	None

From Gerarde, H. W., *Toxicology and Biochemistry of Aromatic Hydrocarbons*, Browning, E., Ed., Elsevier, Amsterdam, 1960, 98. With permission.

between 1941 and 1959. Nearly all of acute benzene toxicity involves inhalation of the benzene fumes. Ingestion either by mistake or in an attempt to commit suicide can occur, but it is rare. Santesson[13] described a woman who drank 30 g of benzene, became comatose, and died 125 hr later. On the other hand, acute benzene toxicity can happen by solvent abuse. Solvent abuse is defined as the intentional or purposeful inhalation of volatile organic chemicals including benzene, toluene, gasoline, etc.[16] As an example of this kind of acute benzene toxicity, we can briefly report the case history of a fatal patient observed recently in Istanbul. A 16-year-old boy was a worker in a small plant where a solution containing 94% benzene was used for cleaning lockers. The boy was accustomed to sniffing this solution. On the accident day, he sniffed pure benzene liquid in an isolated part of the plant and suddenly died while sniffing. The pathologic data of this case are briefly described in the pathology section. A slightly different case of fatal benzene toxicity by benzene abuse was reported by Winek and Collom[17] in the U.S. An 18-year-old boy was found dead after sniffing reagent-grade benzene from a plastic bag in his locked college dormitory room. Autopsy showed a massive hemorrhaging of the lungs, infarction of the spleen, acute congestion of the kidneys, and marked cerebral edema. The content of benzene in the blood was 2 mg/100 mℓ, and it was found between 0.06 and 3.9 mg/100 mℓ in different tissues.

C. Symptoms

The clinical findings, the severity of symptoms, and outcome of acute benzene toxicity depend on the amount of toxic vapor inhaled and the duration of the exposure. The first symptoms are mostly related to the central nervous system. In severe cases, unconsciousness, convulsion, delirium, salivation, nystagmus, phenomena of very intense asphyxiation due to paralysis of the medullary respiratory center, and sudden death will occur. Inhalation of high concentrations of benzene may have an initial stimulatory effect on the central nervous system characterized by exhilaration, nervous excitation and or gridness, nausea, incoherent speech, and headache, followed by a period of depression, drowsiness, fatigue, or vertigo; dermatitis may occur. The benzene odor on the breath of the patient, as well as the flushed face, can be helpful in diagnosis. Although the face will look pink, it is not cyanotic. Cardiac arythmia appears early. There may be a sensation of tightness in the chest accompanied by dyspnea, and ultimately the person loses consciousness, convulsions occur, and finally sudden fatal cardiac arythmia appears. Following the crises, headache, ringing of the ears, nystagmus, spastic paresis, heart trouble, dyspnea, weakness, digestive system disturbances, exhaustion, insomnia, drowsiness and genital complaints might be reported.[17-20] Aspiration of a small amount of liquid benzene immediately causes pulmonary edema and hemorrhaging of pulmonary tissue. If 30 g of benzene is taken orally, the result may be fatal.[13,18] Furthermore, the severity of the symptoms of acute benzene toxicity depends, except for concentration and the duration of exposure, on individual susceptibility. Browning[19] reported a case where the person attempting to rescue a worker overcome by benzene in the benzene tank died, while the original victim survived. The adrenal glands have been implicated in the context of individual susceptibility; or a possible deficiency of an unknown enzyme involved in the detoxification of benzene may be responsible. On the other hand, muscular exertion and fear are believed to augment the severity of intoxication of benzene. Sometimes in acute benzene toxicity, instant or sudden death can occur, as evidenced in Täuber's[20] case.

Laboratory findings in acute benzene toxicity: in general, blood samples are dark red. Leukocytosis and sometimes even polycythemia might be present. In most cases very little benzene will be detected in the urine such as 0.06 mg/100 mℓ. An evaluation in conjugated etheral sulfates and phenol derivates in urine that also increase in chronic benzene toxicity may be helpful. In addition, benzene can be detected in blood, urine, and in different tissues. Fatal cases of sniffing death showed high concentration of benzene in the brain (3.9 mg/100 mℓ) and in the fat (2.2 mg/100 mℓ).[17]

D. Pathology

Animal experiments give the impression that acute benzene toxicity is accompanied by a functional hypertrophy in the cortex and medulla of the adrenal glands.[19,23] A rapid and excessive hormone secretion into the circulatory system seems to take place. In human beings, the results of acute benzene toxicity are known from autopsies performed in such cases. All the organs including the brain, digestive system, pleura, lungs, pericardium, kidneys, bladder, and urether show evidences of hemorrhage. There are several petechial hemorrhages of epidermal tissues. Kidney congestion and cerebral edema are prominent. All internal organs exhibit hyperemia. Even cyanosis of the liver and spleen may be present. In some cases certain organs, like the lungs, will have the odor of benzene.

If benzene is determined in the various tissues, the highest concentration of benzene will be found in the blood, followed by fatty tissue and medulla. The blood is dark, cyanotic red. As an illustrative example of the pathologic findings of acute benzene toxicity, we shall summarize autopsy findings of a worker with acute benzene toxicity who had worked in Istanbul in a plant where this chemical was used.[24] This case was mentioned above.

"There were scattered hyperemic or hemorrhagic spots in the dermal region of the head. There was hyperemia in the meninx, and brain was oedematous. There were small bleedings in meninx.[24] In thorax both lungs were oedematous and there were diffuse pinpoint bleedings. The organs of the abdomen were hyperemic. The mucous membranes and serosa were also hyperemic. The blood obtained in autopsy was analyzed and found to contain alcohol 20 mg/100 mℓ (normal: 30 mg/100 mℓ). There was no carbon monoxide. The pieces of abdominal organs did not exhibit narcotics and other toxic materials. The liquid material which was used as thinner in the workplace contained 94% benzene."

E. Late Effects of Acute Benzene Toxicity

There are only a few late effects of acute benzene toxicity and they are not usually serious. However, it should be remembered that an individual with acute benzene toxicity became edematous and died 2 months following complete recovery.[19] The symptoms of late effects of acute benzene toxicity are nervousness, depression, insomnia, staggering gait, paleness, tachycardia, respiratory catarrh, and pleurisy, as well as cardiac distress. Hepatorenal syndrome may also occur. Derot and Philbert[21] have reported hepatorenal syndrome that led to coma in a girl who had ingested benzene by mistake 2 months before.

F. The Effect of Benzene on Eyes

Effects of benzene vapor or liquid on the eye are negligible; but very high concentrations may cause some smarting sensation in the eye. Direct contact with benzene may cause erythema or blistering. On the other hand, droplet contamination of the eye by benzene causes a moderate burning sensation, but this is only a slight, transient injury of the epithelial cell of the eye which recovers rapidly.[25]

G. Pathogenesis

According to Winek and Collum,[17,18] the mechanism of acute benzene toxicity is very obvious. As it is known, benzene is lipid soluble. When benzene is inhaled in high concentrations, it will accumulate in the organs depending on their lipid content.[17,18] Therefore, benzene accumulates in the brain and body fat and acts as a typical narcotic poison. On the other hand, benzene is considered a pharmacologic narcotic but not a legal narcotic.[17] According to Moeschlin,[18] acute benzene toxicity corresponds completely to acute gasoline poisoning; deep anesthesia develops and if the person is not removed from the environment of benzene, a rapid respiratory or cardiac arrest will ensue. For the sudden death of acute benzene toxicity, primarily asphyxiation or the development of ventricular fibrillation are accused.[17] On the other hand, the rapid discharge of adrenalin in circulation is also taken into consideration in sudden death which happens in some cases of acute benzene toxicity.[17]

H. Prevention

Means of protection against the toxic effects of benzene will not be discussed here. Only precautions to avoid acute benzene toxicity will be mentioned. If possible, benzene storage tanks should be discharged and cleaned by mechanical means. As Infante and White[26] emphasized, even in 1980, fatal acute benzene toxicity occurred following entering a benzene tank. "A 40-year-old black male in Baltimore, Maryland, died from acute benzene intoxication after entering a tank where benzene fumes were known to be present.[26] The worker obviously was not afforded appropriate respiratory protection. Permanent workers at the plant refused to enter the storage tank because of the known hazards of this chemical exposure. As a result, this worker was hired from a temporary employment agency to enter this tank. He died after a few hours of exposure to benzene."

I. Treatment

The patient should be immediately taken out to fresh air. If breathing has stopped, artificial respiration or recirculation should be started and oxygen should be administered. Hot water bottles may be placed around arms and legs. Analeptics may be used, but adrenal hormones should be avoided since adrenaline will increase the effects of benzene on the myocardium and may cause fibrillation. The patient should not return to this job for a period of time after recovery.

1. Modulation of Benzene Toxicity by an Interferon Inducer (6MFA)*

Pandya et al. administered benzene (1 mℓ/kg body weight) for 3 days in female albino rats.[26a] This trial produced leukopenia, lymphocytopenia, and increased the number of nucleated cells in the bone marrow and significantly decreased the weight of thymus and spleen in animals. Iron content, lipid peroxidation, and superoxide dismutase activity of the liver and bone marrow were significantly increased as a result of benzene exposure. Prior administration of 6MFA, an interferon inducer with immunomodulating potential, was found to ameliorate some adverse effects of benzene as well as restoration of hepatic architecture histologically. Lipid peroxidation and iron content were both normalized, whereas superoxide dismutase activity was further increased and the number of lymphocytes and bone marrow cells returned to normal. Pretreatment of albino rats with interferon inducer (6MFA) was able to enhance sheep red blood cell antibody titer in benzene-treated, immunosuppressed

---

* 6MFA, 6th myecelial fraction acronymed 6MFA, given to the interferon-inducing antiviral agent, isolated from the fungus *Aspergillus ochraceus*, ATCC 28706.

animals. According to Pandya et al., the beneficial effects of 6MFA in amelioration of acute benzene toxicity deserves further attention.[26a]

II. CHRONIC BENZENE TOXICITY

A. History

Benzene came into use as a solvent for rubber at the end of 19th century, after which cases of aplastic anemia started to occur. In 1888 when bicycle tires became available, benzene-containing adhesives began to be used in the industry.[3] In 1897, Santesson,[13] Professor of pharmacology in Stockholm, reported nine cases with chronic benzene toxicity. Four female workers died from aplastic anemia in Upsala, Sweden. Afterwards, Selling[27,28] was the first to describe three cases of chronic benzene toxicity, two terminated fatally in the U.S.[29] These cases occurred among the workers in the coating room of a cannery plant in Baltimore, Md. where 14 young girls were employed. The concentration of benzene in the air of the workroom was apparently sufficient to produce fatal cases of chronic benzene toxicity. Case reports and autopsy findings in the two patients were consistent with the diagnosis of fatal aplastic anemia due to chronic exposure to benzene. The third female worker showed a mild anemia associated with borderline normal leukocyte count. The patient recovered rapidly following admission to hospital.

Selling[28] emphasized the toxic effects of benzene on the granulocytic series as a powerful leukotoxin and proposed that hemorrhagic diathesis observed in the cases of benzene toxicity was a secondary effect. Selling also stressed that benzene attacks the entire hematopoietic or blood-forming system in general. He ascribed this to aplasia in the bone marrow. After Selling demonstrated the leukopenic effect of benzene, Koranyi[30] tried it in the treatment of leukemia. Koranyi administered a daily dose of 3 to 5 g of benzene orally to his leukemic patients. Although leukocytes initially decreased and patients improved somewhat, they later developed fatal aplastic anemia. Similarly, Rochner et al.,[31] in the U.S., used benzene in the therapy of chronic myeloid leukemia. Koranyi also used small doses of benzene (0.25 g) to stimulate erythropoiesis.[30] In 1930, Schneider[32] demonstrated that the hemorrhagic diathesis observed in benzene-mediated aplastic anemia was due to secondary thrombocytopenia. Hunter[3] was right to point out that following Selling's studies on benzene, a conception of a simple constant clinical picture was established; the attack of benzene on the bone marrow was destructive, affecting first the platelets, then the granulocytes, and finally the red cells. According to this assumption, diagnosis is not justified unless the blood picture shows the findings of aplastic anemia such as granulocytopenia, thrombocytopenia associated with purpuric manifestations, absence of splenomegaly, and the presence of aplasia of bone marrow at necropsy.

In 1918 in England, xylene was substituted for benzene wherever possible in an effort to diminish the victims of aplastic anemia due to chronic exposure to benzene since the frequency of such cases was increasing.[3] Alice Hamilton,[33] realizing the hazards of benzene, started a campaign to increase the safety measures in the areas of benzene use.[3] The efforts were successful and resulted in the substitution of toluene and petroleum naphtha for benzene in industry. With the introduction of rubber latex, a product of the rubber tree, as Hunter emphasized,[3] another big step away from benzene became possible. When photogravure was invented in 1933, benzene-mediated health problems in England and the U.S. increased because dyes used in the photogravure contained 10 to 15% benzene, and exposed workers in this field to its hazard.[3] Greenburg et al.[34] performed an investigation among 332 postgravure workers and they found in 130 of them a variety of hematological abnormalities characteristic for chronic benzene toxicity. As proper ventilation began to be provided in the areas of benzene use, and the other solvents began to be substituted for benzene, health problems associated with solvents decreased.[3]

In Turkey, prior to 1955 to 1961, benzene was only infrequently used as a solvent by shoe workers. However, it seems very probable that this aromatic hydrocarbon has been used even rarely as a dilutent in other workplaces. For instance, Prof. E. Frank[35] presented a whitewasher with aplastic anemia, possibly due to benzene exposure. At that time the shoe workers in Istanbul and in the other cities prepared their glue adhesives by dissolving rubber in gasoline.[36] As they learned that glue adhesives dissolved in benzene were extremely practical and much cheaper, they quickly replaced their old method with this new product. After 1961, the number of patients with benzene-induced aplastic anemia gradually increased and reached 52 in 1978.[37,37b] After a phase out of benzene in Istanbul and the other cities, the number of the victims of chronic benzene toxicity gradually decreased in Turkey.[38]

B. The Frequency

To determine the exact frequency of benzene toxicity in a working place or in a population is not easy and is sometimes controversial. The screening procedures for hematological abnormalities and the interpretation for this purpose vary greatly. As Goldstein pointed out,[39] the literature reporting benzene hematotoxicity in humans falls into two categories: (1) single case reports or series of case reports associated with chronic exposure to benzene and (2) true incidence studies showing benzene induced hematological abnormalities in a workplace or population who are occasionally exposed to benzene.

Precautions introduced in the 20th century in Europe and the U.S. have radically decreased the number of cases due to chronic benzene toxicity. The reasons for this decrease are the following:[3]

1. Whenever possible, the use of less toxic solvents such as toluene, xylene, saturated or aliphatic hydrocarbons, and other solvents selected according to the nature of the substance to be dissolved, alcohol, ether, etc.
2. More stringent monitoring of environmental contamination.
3. Prevention of benzene leaks.
4. Improved working conditions including more stringent medical surveillance.

Despite these precautions, cases of benzene-induced aplastic anemia and leukemia are still encountered everywhere. Examples are numerous. In 1956, Savilahti[40] encountered chronic benzene toxicity in a large group of workers in a shoe factory in Finland. Lob[41] found cases of chronic benzene toxicity at a gluing unit of a paper factory in Switzerland. One of them showed the findings of acute leukemia. In Japan in 1960, 50% of the employees of workshops that utilized benzene, exhibited hematological abnormalities and cases of aplastic anemia.[42] Furthermore, organic solvents including benzene were suspected again as causative factors in 6 to 12% of the patients with aplastic anemia.[43] In his testimony to OSHA, Sakol,[44] an Akron hematologist, described what he considered to be an epidemic of a rare form of leukemia among certain workers. From 1954 to 1963 he observed nine cases including at least four and probably nine of erythroleukemia; all were possibly exposed to benzene during a pliofilm operation in Ohio. Recently, in a study in China, it was established that 528,729 workers were exposed to benzene or benzene mixture in 28 provinces of China during the period of 1979 to 1981.[45,45a] There were 27,808 factories using benzene as a solvent or as a chemical intermediate. Among the workers examined, 2676 cases of benzene toxicity were found. The prevalence rate was 0.51%. Among benzene-induced hematological abnormalities, 24 cases of aplastic anemia and 9 cases of leukemia were found. This study showed that cases of benzene poisoning may occur, even in factories with low levels (<40 mg/m^3 = 11 ppm) of concentration.[45]

On the other hand, there are still reports on the occurrence of chronic benzene toxicity including leukemia among individuals who were exposed to benzene for nonindustrial purposes. In 1977, Brandt et al.[46] described four cases of acute leukemia due to nonindustrial use of benzene in Sweden. Yet, benzene-induced hematological disorders have been practically eliminated in countries where precautions to prevent undue exposure to benzene are in effect. Browning[19] reported that she found only four fatal cases in England in the last 18 years. Hunter[3] emphasized that no hematological problems were found in 1200 women who had to utilize benzene in England during World War II. This conclusion was reached after complete hematological examination. In his book, Hunter ascribed the absence of hematological problems to the precautions taken to prevent undue exposure to benzene and congratulated the staff of the factory department of the Ministry of Labor for the way in which they protected the worker from the hazard.[3] Today benzene-induced aplastic anemia is a rare phenomenon in Western countries due to preventative measures and medical surveillance. Contrary to this, in Eastern countries including Turkey, possibly the most important causal agent for the development of aplastic anemia is benzene exposure due to insufficient preventive measures, precautions, and medical surveillance in workplaces.[43,47] As an example, in 1983, in order to illustrate the etiologic role of drugs and chemicals in the development of aplastic anemia in Istanbul, we analyzed 108 cases of aplastic anemia among 3715 hematologic patients during a period of 10 years (1973 to 1982) at the hematology section of Istanbul Medical School.[47] In 52 (48.1%) of 108 patients, a possible etiologic factor was found. In contrast to data obtained in Western countries, the most important causal agent for the development of aplastic anemia in this series was benzene, which accounted for 25 (23.1%) of the cases. These results are a little disappointing because the use of benzene as a solvent has been officially prohibited since 1969 in Turkey.

C. Factors that Influence the Frequency of Benzene Toxicity
1. Individual Susceptibility

Individual susceptibility is one of the most important variables in the development of chronic benzene toxicity. Unfortunately, at present there is no test to evaluate the susceptibility of an individual to the toxic effect of benzene. As early as 1922, Reifschneider[48] and Ronchetti,[49] and later Hamilton,[33] suggested family susceptibility to this chemical as one of the main factors in the frequency of chronic benzene toxicity. Reifschneider reported on a family with three cases of benzene toxicity, two of which were fatal. We can give some examples from our series of chronic benzene toxicity in support of this assumption. Two of our thalassemic patients with aplastic anemia due to chronic benzene toxicity, one fatal, were brothers.[50] Also two patients with pancytopenia associated with chronic exposure to benzene, one fatal, were cousins.[51] Similarly, the son of one of our patients with fatal aplastic anemia due to chronic benzene toxicity showed leukopenia following a short time after exposure to this chemical.[52] As a very illustrative example of individual susceptibility, we should like to present here a family associated with chronic exposure to benzene. A 38-year-old man and his 32-year-old wife were engaged in manufacturing whistles. For this purpose they dipped plastic material into an open vessel of benzene solution containing 88.42% benzene and 9.25% toluene, as determined by gas chromatography. Following exposure to benzene for a period of 6 months, a severe aplastic anemia developed in the wife.[53,54] Contrarily, in her husband, following an exposure of 14 years, all hematologic data were within normal limits.[54] The cause of this variation concerning individual susceptibility is not completely understood. According to Browning[19] the most probable explanation lies in innate difference in the potency of different individuals to carry out the metabolic detoxification. Some individuals may be congenitally deficient in the capacity to detoxify benzene or its metabolites. On the other hand, there are some investigators who do not accept individual susceptibility. According to Goldstein,[55] "Although there is some suggestion of differences in the extent of susceptibility to benzene, perhaps related to hormonal factors, body build and genetic disposition, it does appear that all humans are susceptible to the pancytopenic effects of this agent." Despite this, we are strongly inclined to believe in the important role of individual susceptibility in the development of chronic benzene toxicity.[54]

2. Age and Sex

The opinion that women and young people are especially susceptible to benzene toxicity was held by many of the earlier investigators dealing with this subject, but it is less firmly believed today. However, some observations might indicate that more cases of benzene toxicity are seen in women than in men. Hirokowa[56] reached this conclusion in Japan. He performed experiments in rabbits and showed that female rabbits and male rabbits casto-rectemized and injected with ostradiol had comparatively lower levels of leukocytes and erythrocytes following benzene injections than did male rabbits. Ito,[57] in Japan, pointed out that benzene toxicity depends on the hormonal influence.

Contrary to these data, Siou et al.[58] obtained opposite results in the sex difference relating to the mutagenic effects of benzene. For this purpose, they applied a micronucleus test to mice aged 21 days and adult mice aged 5 weeks. The results showed that there is no difference between male and female mice at 21 days, but that from an age of 5 weeks, males are approximately twice as sensitive as female mice to the mutagenic effect of benzene. Furthermore, Siou et al.[58] showed that castrated males were as sensitive as females, but regained their original sensitivity when treated with testosterone. Gad-El-Karim et al.[59,60] had similar results with micronucleus tests and metaphase analyses, namely, female mice were consistently more resistant to benzene than males. On the other hand, the sex difference in the susceptibility to hematopoietic disorders induced by benzene was studied kinetically by Sato et al.[61] with a special reference to its relation with the body fat content. In rats of both sexes with a large body fat content, benzene was eliminated more slowly and remained in the body for a longer time than in rats with a small body fat content. According to Sato et al.,[61] the decrease in leukocytes during chronic benzene exposure was observed only in the groups of rates associated with a large volume of fat tissue. In an experimental human exposure, the elimination of benzene was slower in females than in the males. The kinetic study showed that the slower elimination in the females is primarily due to the bulky distribution of body fat tissue in that sex. From these animal and human experiments, Sato et al.[61] concluded that females, with their massive body fat tissue, show an inheritent disposition to be susceptible to benzene which has a high affinity with fat tissue.

At one time, no relationship between age and benzene sensitivity was thought to exist. However, the experts that met in Geneva[62] between May 16 and 22, 1967 accepted a report that suggested individuals from both sexes under 18 years of age should not hold jobs that would involve exposure to this hazard. This was based on the conclusion of hematologists who indicated that adolescents are more sensitive to substances that have a toxic effect on the bone marrow. On the other hand, Doskin[63] in the U.S.S.R., performed a hematologic study on 365 workers and technicians of different age groups exposed to a combination of hydrocarbons in which benzene was the main effective agent. This study showed that the highest incidence of hematological alterations occurred among the workers and laboratory technicians who had started to work at the age of 18 to 21. Considering these data, Doskin recommended that measures for workers up to 21 years of age exposed to benzene combined with other hydrocarbons should be primarily undertaken.

Contrarily, the present author and associates[64] failed to show a correlation between age and increased susceptibility to benzene-induced hematological abnormalities; 45 out of 217 individuals studied were in the 13 to 20 age bracket. Similarly, 23 out of 51 workers with hematological abnormalities associated with chronic benzene exposure were in the 12 to 20 age bracket. On the other hand, the present author and Erdem,[37] in a follow-up study on 44 pancytopenic patients with chronic benzene toxicity, showed that there was a statistically significant relationship between the age of the patients and the mortality, and the development of leukemia.

D. Relation to Duration of Benzene Exposure

Correlation between the occurrence and the severity of hematological abnormalities and

Table 2
EVOLUTION OF RECOMMENDED CONTROL
LEVELS IN THE U.S.

1920s	100 ppm for 8-hr day (Mass. and some other states)
1930s	50 ppm for 8-hr day (state standard)
1940	35 ppm for 8-hr day (state standard)
1942—1945	100 ppm for 8-hr day Federal War Time Standard
1946	100 ppm ACGIH[a] — TLV[b]
1947	50 ppm ACGIH — TLV
1948	35 ppm ACGIH — TLV
1957	25 ppm TWA — ACGIH — TLV
1963	25 ppm ceiling value — ACGIH — TLV
1969	10 ppm TWA with 25 ppm ceiling — ANSI[c]
1974	10 ppm with 25 ppm ceiling with 5 ppm action level — NIOSH[d]

[a] American Conference of Governmental Industrial Hygienists Limit.
[b] Threshold limit value.
[c] American National Standard Institute.
[d] National Institute Occupational Safety and Health.

From Weaver, N. K., Gibson, R. L., and Smith, L. W., *Advances in Modern Environmental Toxicology, Carcinogenicity and Toxicity of Benzene*, Vol. 4, Mehlman, M. A., Ed., Princeton Scientific Publishers, Princeton, N.J., 1983, 63. With permission.

the duration of exposure to benzene varies widely. Chronic benzene toxicity can occur without lengthy exposure to benzene. According to Savilahti,[40] if benzene concentration in the working environment is high enough, 2 months of exposure can result in toxicity. Similarly, one of 44 patients with benzene-induced aplastic anemia in whom preleukemia developed 6 years later, aplastic anemia had resulted from exposure over a 4-month period.[37,65] In the study of 217 workers, we found that hematological abnormalities occurred in about one tenth of them within the first year of exposure.[64] It is obvious that as the duration of exposure to benzene increases, the ratio of chronic benzene toxicity cases will increase also. On the other hand, present author and Erdem,[37] in a follow-up study reviewing the mortality and the development of leukemia in 44 pancytopenic patients, established that there was a statistically significant relationship between the duration of exposure and the development of mortality and also that of leukemia.

E. The Concentration of Benzene in the Workplace and the Development of Hematological Abnormalities

Unfortunately, exact and reliable data showing the correlation between the concentration of benzene in the air of workplaces and the occurrence of hematological abnormalities including leukemia are lacking.[23] On the other hand, there are no reliable data showing the absence of hematopoietic disorders in workplaces with a low level of benzene concentration in the air. When the concentration of benzene in a workplace increases, particularly above 50 ppm, the hematological abnormalities including aplastic anemia and leukemia appear rapidly. This fact was observed in several countries.[22,37,65] Therefore, the recommended control levels in workplaces in the U.S. dropped from 100 to 10 ppm. These data are summarized in Table 2. In 1977, the OSHA attempted to lower the exposure limit from 10 to 1 ppm,[66,67] but the Supreme Court, in a five to four decision, rejected the OSHA ruling. This decision evoked a continuous discussion.[68-72]

According to the present author, the permissible limit for benzene should be 1 ppm; where it is technically impossible to achieve this, the exposure limit should be as low as possible. The reasons proposed for this decision are as follows:[67-69]

1. The leukemogenic and carcinogenic effects of benzene have been shown unequivocally.
2. Studies showing the absence of hematological abnormalities in the individuals with occupational exposure in very low concentration are unsatisfactory.[67-70,73,74]
3. The great variation in the individual susceptibility; this phenomenon has been considered as one of the main factors in the incidence of chronic benzene toxicity.
4. Several studies have shown that the duration of exposure to benzene and the subsequent development of bone marrow abnormalities, including leukemia, varies widely. In our series, the duration of exposure in aplastic anemia varied from 4 months to 15 years.[37,65] In the leukemic patients with exposure to benzene, the duration of exposure varied from 4 months to 40 years (Chapter 6).
5. Some investigators were able to show important abnormalities in porphyrine metabolism and a decrease in leukocyte alkaline phosphatase activity in apparently healthy workers with normal hematological findings who were chronically exposed to even moderate or low levels of benzene, such as 10 and 25 ppm.[75,76]
6. In workers chronically exposed to benzene of 5 to 25 ppm, despite the absence of hematological findings, there were significant increases in the frequency of chromosomal aberrations from blood lymphocytes.[77,78]
7. As will be explained in Chapter 6, in a mortality experience of 594 workers exposed to benzene in the chemical industry performed by Ott et al.,[79] three patients' deaths were due to acute leukemia. The time-weighted average benzene exposure of these leukemic individuals was below 10 ppm.
8. Baarson et al.[80] performed experiments in which mice were exposed to 10 ppm benzene for 6 hr/day, 5 days/week for 178 days. This trial caused in mice a progressive depression in the in vitro colony-forming ability of one of the erythroid progenitor cells, the colony-forming unit-erythroid (CFU-E). Colony growth of cells from exposed mice was only 5% of the control colony growth after 178 days of exposure. Burst-forming cell growth was depressed to 55% of the control group after 66 days, but returned to control growth values at 178 days. In addition, benzene-exposed mice exhibited depressions in the numbers of splenic nucleated red cells and in the numbers of circulating red cells and lymphocytes. According to Baarson et al.,[80] the results mentioned above suggest that low level exposure to benzene may be hematoxic.
9. Chang[81] in Korea, performed a hematological study in the workers exposed to benzene. According to Chang, the results obtained were as follows: (1) even though some workers were assumed to be exposed to 20 ppm of benzene, they showed anemia and leukopenia after 42 and 96 months of working duration; (2) according to Chang, with regard to the hematological changes, the following relationship could be formulated between the concentration of benzene in air (y = ppm) and the duration of working (x = months) y = 82.5x 0.77 0.2x + 10.1; and (3) it is presumed theoretically that the benzene toxicity may occur when workers are exposed to as low as 10.1 ppm benzene in air.

Considering the above-mentioned data we are inclined to accept that the permissible level of benzene in the working environment should be 1 ppm. Where it is technically impossible to achieve this, as Truhaut[82] suggested and practiced in Sweden, a temporary limit of 5 ppm would be acceptable, but with the condition that it will drop to 1 ppm later.[83,83a]

F. Pregnancy and Abortion
The experts of the meeting on the toxicity of benzene, held in Geneva-ILO (International Labor Office), May 16 to 22, 1967, concluded that pregnant women and nursing mothers

should not be exposed to the hazards of benzene.[62] Since pregnancy is not immediately evident, when woman of childbearing age are employed in areas which risk exposure to benzene, special care is necessary. Lachnit and Reimer[84] proposed that pregnancy potentiates benzene toxicity. They cite a female who worked 10 years in the rubber industry and developed a severe aplastic anemia 2 years after she got married. The outcome was fatal. Another female case who worked with an adhesive containing xylene and supposedly toluene for 2 years, became pregnant and developed fatal aplastic anemia.[84] Saito and Perini[85] reported a mother that had an abortion and died from benzene-mediated aplastic anemia; 3 1/2 years later, one of her children died of benzene-mediated aplastic anemia. Two cases reported by Savilahti[40] were pregnant and exhibited all findings of aplastic anemia. In both cases, the pregnancies continued normally, and the children developed normally. In contrast, in some animal experiments the results were opposite, namely, the leukopenic effect of benzene was ameliorated in pregnancy possibly as a result of accelerated benzene detoxification due to increased steroid metabolism.[86]

G. The Relationship Between Infections, Leukocyte Anomalies, and Chronic Benzene Toxicity

Browning[19] suggests that infection will increase the individual susceptibility to benzene. Diseases of respiratory organs, especially tuberculosis, diabetes mellitus, heart disease, nervous disorders, nephritis, and obesity are the factors that predispose an individual to chronic benzene toxicity.[19] Three of our patients with aplastic anemia due to chronic exposure to benzene had lung tuberculosis.[87] One of our patients with aplastic anemia due to chronic exposure to benzene had also Pelger-Huet leukocyte anomaly.[87] Despite this, the inborn leukocyte anomaly did not cause any aggravation of benzene toxicity on the hematopoietic system. Koslova[88] conducted a hematological study among 273 workers exposed to benzene. The study showed that during the summer months, the number of workers with hematological abnormalities increased. According to Koslova, experimental studies with laboratory animals disclosed similar results.

H. Ethyl Alcohol and Chronic Benzene Toxicity

There is an abundance of information in the literature on the interactions of ethyl alcohol (ethanol) and industrial chemicals and drugs.[89] One of these chemicals is benzene. We had a clinical impression that in 12 of our patients with pancytopenia or leukemia due to chronic exposure to benzene, the abuse of alcohol increased the toxicity of benzene on these workers.[90] Snyder et al.[91] and Baarson et al.[92] performed animal experiments in which mice were exposed to 300 ppm benzene by inhalation and 5 or 15% of ethanol in drinking water. The results demonstrated that in mice, the combination of inhaled benzene and ingested ethanol was more hematotoxic than inhaled benzene alone. Anemia and lymphopenia were observed in the peripheral blood of all benzene and benzene/ethanol-treated groups. These cytopenias, however, were more severe in the benzene/ethanol-treated mice. In addition, there was a transient increase in normoblasts in the peripheral blood of benzene/ethanol-treated mice which was not observed in mice treated with benzene alone or in the control mice. Compared to treatment with benzene alone, the combined treatments produced a more severe marrow and splenic aplasia. According to Baarson et al.[92] the ingestion of ethanol increases the hematotoxicity of inhaled benzene. Furthermore, Rosenthal and Snyder[93] extended these trials with benzene/ethanol combination in splenectomized mice; splenectomized and spleen-bearing mice were used to explore the source of normoblasts appearing in the peripheral blood of animals treated with benzene and ethanol, and including the role of splenic hematopoiesis in compensating for bone marrow stress by repeated benzene and ethanol combined exposure. Regardless of status, mice exposed to the combined treatment demonstrated a transient appearance of normoblasts in the peripheral blood. According to Rosenthal and Snyder,[93] bone marrow is an apparent source of peripheral normoblasts. This phenomenon was not seen in mice treated only with benzene or ethanol. Splenectomy significantly influenced the bone marrow response to the hematopoietic effects of benzene alone, as well as the combined treatment of benzene and ethanol, as evidenced by altered marrow normoblast and granulocytic equilibria. On the other hand, ethyl alcohol may cause folate deficiency. Therefore, the transient appearance of peripheral normoblasts could partly be explained by this possibility.

Driscoll and Snyder[94] performed another animal experiment in which effects of ethanol and repeated benzene exposure were evaluated. Two groups of mice were exposed to benzene at 3000 ppm, 6 hr/day, 5 days/week for 20 exposures. One group received 10% ethanol commencing 20 hr prior to the initial exposure and continuing 5 days/week throughout the study. The second group received tap water. During the first exposures, the mean steady state (Css) in benzene/ethanol-treated mice and benzene/water-treated mice were 5.2 and 10.7 mg/mℓ, respectively. The mean elimination rate constants for benzene/ethanol and benzene/water-treated groups were 0.124 and 0.042 min^{-1}, respectively. But these data were different after the 20th exposure. According to Driscoll and Snyder,[94] the results obtained in their animal experiments indicate that 1 day of 10% ethanol consumption causes a dramatic effect of benzene kinetics. After 20 days of treatment, the benzene/water and benzene/ethanol animals are kinetically similar. According to Driscoll and Snyder, these changes in kinetics can be explained by the ability of ethanol and benzene to alter benzene metabolism.

I. Thalassemia, Abnormal Hemoglobins, and Chronic Benzene Toxicity

There are few studies showing the possible relationship between thalassemia, abnormal hemoglobins, and exposure to benzene. In 1959 Saita and Moreo[95] have reported three cases of chronic benzene toxicity involving beta-thalassemic heterozygotes. According to Saita and Moreo, beta-thalassemia does not potentiate benzene toxicity. Similarly, Gaultier et al.[96] studied individuals working in benzene-related areas in Paris, but they were unable to demonstrate a difference in the sensitivity of individuals with beta-thalassemia or with other hemoglobinopathies such as heterozygous hemoglobins S and C, as compared to normal individuals. In our series of chronic benzene toxicity, four patients were also beta-thalassemia heterozygotes. Two of them who were brothers, one fatal, showed high levels of fetal hemoglobin, ranging between 30 and 60%, during the period of pancytopenia.[37,50] In the third beta-thalassemic patient, during the course of chronic benzene toxicity, pseudo-Pelger anomaly developed.[64] Following cessation to exposure, this leukocyte anomaly disappeared. On the other hand, sometimes exposure to high levels of benzene can cause subacute hemolytic crisis in the individuals with heterozygous thalassemia.[97] In 1961, Saita and Moreo[97] described a worker who was exposed to a high level of benzene. The solvent used contained 65% benzene. A few days after exposure, a subacute hemolytic crisis developed. There was a decrease in the level of red blood cells and hemoglobins. Following some days, the red cell and hemoglobin contents returned to prehemolytic crisis level and mild bilirubinemia disappeared.

The observation mentioned above, namely, acquired pseudo-Pelger anomaly and very high level of fetal hemoglobin in beta-thalassemic heterozygotes associated with chronic benzene toxicity, led us to suggest that some forms of beta-thalassemia increase the deleterious effects of benzene on the hematopoietic system.[50,64] On the other hand, Simson and Shandlar[98] emphasized the necessity that if beta-thalassemic workers continue their jobs involving exposure to benzene or lead, it should be controlled much more strictly than usual medical and in-plant industrial hygiene surveillance.

J. Symptomatology

Clinical findings in chronic benzene toxicity are quite variable, vague, and nonspecific.

Subjective complaints, such as headache, dizziness, nausea, vomiting, and loss of appetite, etc. may precede the changes in blood counts. Initially, the clinical picture of chronic benzene toxicity was described according to the incomplete description of Santesson[13] and Selling.[27] Briefly, all the symptoms observed in benzene toxicity were ascribed to the effects of benzene on the bone marrow.[27] It was described as first causing the arrest of the production of platelets, followed by the inhibition of granulocyte production, with erythropoiesis ceasing last. These in turn led to aplastic anemia and granulocytopenia. Leukopenia was considered as being the most important symptom, aplastic anemia being secondary, and the symptoms of hemorrhagic diathesis due to thrombocytopenia was thought to accompany all cases of chronic benzene toxicity. In necropsy, the bone marrow is always aplastic. Alice Hamilton[33] was the first to object to such an oversimplification of the clinical and hematological picture of chronic benzene toxicity. She proposed that chronic benzene toxicity, like radium poisoning or X-ray sickness, could exhibit variable clinical and hematological findings as well as those of the bone marrow picture.[33] The data that has accumulated on chronic benzene toxicity since 1939 has shown that Hamilton was right.[3] In 1913, Koranyi[30] had demonstrated that erythropoiesis will be initially stimulated by a low dose of benzene.

Unfortunately, the onset and progress of chronic benzene toxicity is very elusive. The patient and cohort realize the situation very late, at about the time the bone marrow has been affected and a doctor is seen. As a matter of fact, we found 5 out of 44 aplastic anemia cases while screening workers that were exposed to benzene at their jobs.[37,64,87] These workers told of their health problems only after they were admitted to the clinic for observation. The educational level of the patient is very significant in this regard. There is a causal aloofness towards health, and health hazards arise when the educational level is low. This also increases the delay in medical attention, resulting in an increase in the number of benzene toxicity victims. Similarly, Greenburg et al.[34] mentioned among 312 workers exposed to benzene, five patients with mild or moderate pancytopenia who made no complaints and were negative on physical examination. According to Greenburg et al., "Evidently benzene poisoning may be present without the individual being aware that anything is wrong."[34]

Certain changes can be the first indications of chronic benzene toxicity. Browning[19] lists them as follows: dryness of skin and mucous membrane due to fat-dissolving action of benzene, skin irritation, dermatitis of eczematous type, rash, fatigue, insomnia, lethargy, and sometimes purpuric spots of the mucous membrane and epistaxis may be observed. Nausea and burning sensation in the epigastrium and sometimes vomiting are the digestive tract related symptoms that can serve as early indications of benzene toxicity. In addition to these, some individuals exposed to benzene complain of dizziness. It appears mostly at the end of a shift. Some feel coldness, headache, blurring in the eyes, and dyspnea following exertion.

In Monaenkova and Zorina's[99] hemodynamic study on the status of the circulatory system in 300 individuals with chronic benzene toxicity, it was shown that in this chemical poisoning, cardiovascular disorders are of a functional nature, are reversible, and are being caused by hematopoietic disturbances and particularly by the severity of anemia.

On the other hand, the patients of Godrot et al.[100] had gastritis and hyperacidity. Individuals that suffer from benzene toxicity will be easily fatigued and, although they suffer from insomnia at night, they will be prone to sleep on the job. According to Mallory et al.,[101] individuals suffering from chronic benzene toxicity gain weight rather than losing it, while Browning[19] claims that there are no changes in weight. Before chronic benzene toxicity causes pancytopenia, more correctly called aplastic anemia, it may cause other hematological changes, rather mildly. There are many such alterations and there is no agreement as to the sequence and frequency in which they occur. We will discuss them one by one.

K. Leukopenia

After Selling,[27,28] leukopenia was considered to be the first indication of chronic benzene toxicity. Later this belief fell into disregard. Hamilton[33] suggested that leukopenia might not always be the first indication of chronic benzene toxicity. She further suggested that serious bone marrow degeneration can occur in response to chronic exposure to benzene, without leukopenia, and urged to give importance to erythropoietic series that erythrocytes be counted along with the leukocyte. Goldwater[102] reported leukopenia in only 14.5% of 332 employees of printing shops that he studied. This value was less than the extent of anemia, macrocytosis, and thrombocytopenia that he detected. Savilahti[40] found leukopenia in 32.5% of the 147 workers studied, a ratio that was less than those exhibiting thrombocytopenia and anemia. On the other hand, the present author and colleagues found that leukopenia was the most frequently encountered hematological alteration in the 217 workers studied[64] (leukopenia alone 9.72%; leukopenia plus other such as thrombocytopenia and pancytopenia 17%). Hamilton-Paterson and Browning,[103] in a study involving 200 women working with benzene-containing materials, determined that the neutropenia and leukopenia were the most frequently encountered hematological alterations. Studies performed on these individuals after they were away from their jobs for 3 months indicated a complete return to normality. Danysz[104] showed a sensitization pattern in two women that exhibited neutropenia. The pattern took shape; the woman started to manifest hematological changes after returning to work, following a period of furlough. The sensitization pattern was not observed in men. Bernard-Pichon[105] studied 350 workers and calculated that neutropenia was the earliest abnormality, but contrary to the above study, a relapse of neutropenia was not found in the workers that returned from a furlough.

L. Leukocytosis

Although leukocytosis accompanies some cases of chronic benzene toxicity, in general it has not been possible to link this to benzene directly. Goldwater[102] found leukocytosis in the workers he studied. Although it was concluded that chronic benzene toxicity causes leukocytosis, the small number of the cases involved in this study made the conclusion tenuous. Savilahti[40] found leukocytosis in one case, but since the patient was also pregnant, a direct correlation between chronic benzene toxicity and leukocytosis was not possible. We encountered five cases of mild leukocytosis corresponding to 2.31% of the 217 workers tested.[64] If the two cases in which leukocytosis was accompanied by thrombocytosis were added to these five, the total ratio would become 3.23%. However, the changes in two of the seven were due to acute polyarticular rheumatic fever, and two of the remaining cases recovered from leukocytosis quickly. No causes, other than chronic benzene toxicity, could be found in the three remaining cases. On the other hand, Doskin[63] performed a study in the U.S.S.R. on 277 workers and 88 laboratory technicians exposed to the combined effects of benzene, cyclohexane, and divinyl. The first hematologic abnormalities were thrombocytopenia and later moderate normochromic anemia. According to Doskin, the changes in white blood cells, which as a rule commences with leukocytosis (16.5 to 18.5 $\times$ 10^9/ℓ), is followed by leukopenia in some cases. In addition, relative neutropenia due to an increased number of lymphocytosis was observed.

On the other hand, for the present author, leukocytosis, even as an early finding, is not an unusual finding in chronic benzene toxicity. If one considers the fact that several chemotherapeutic agents, such as nitrogen mustard, etc., may cause early and transient leukocytosis during chemotherapy, the appearance of an increase in leukocyte count at the early stage of chronic benzene toxicity is not unusual.

M. Shift to Left

The rod-shaped young granulocytes may increase and are rarely the first indication of chronic benzene toxicity. In our study on 217 workers exposed chronically to benzene, the only finding of chronic benzene toxicity was the presence of pseudo-Pelger anomaly in a worker with beta-thalassemic heterozygote.[64] Except for the findings which usually accompany heterozygous beta-thalassemia, such as mild increase of hemoglobins A_2 and F, hypochromia, and microcytosis, other hematological data were normal. But the majority of the granulocytes had lobe number one or two. When exposure to benzene was eliminated, these anomalies disappeared. This patient is the first case of benzene-mediated pseudo-Pelger anomaly found in a severe benzene toxicity episode. In contrast to our case, in Paterni and Russo's[106] and Saita and Moreo's[95,107] patients, all findings of pancytopenia were present. Furthermore, Zini and Alessandri[108] reported a case of pseudo-Pelger anomaly due to chronic exposure to benzene which was diagnosed 31 months before the death of the patient from acute myeloblastic leukemia. A 37-year-old worker in a shoe factory was exposed chronically to benzene-containing material. A moderate pancytopenia associated with a frank Pelger anomaly was developed. Bone marrow findings showed a relative prevalence of basophilic erythroblasts, myelocytes, and rare megacaryocytes. There were numerous plasma cells. All elements in the granulocytic series showed the character of Pelger-anomaly. The diagnoses pseudo-Pelger anomaly and acute myeloblastic leukemia developed 25 months later.

N. Other Changes of Leukocytes

Although certain findings relating to this subject, such as leukocytosis, are still controversial, important changes in leukocytes are found in some cases of chronic benzene toxicity. These are described below.

1. Lymphopenia

Goldwater[102] was the first to draw attention to lymphopenia in humans during the course of chronic benzene toxicity. Advanced cases of both severe or mild types of chronic benzene toxicity will exhibit relative or absolute lymphopenia. The present author and colleagues found absolute lymphopenia in 28 out of 37 workers with leukopenia (75%).[64] Benzene effects the lymphoid tissue as well as the bone marrow. Latta and Davies[109] showed in rats that benzene affects lymphoid tissue, and they concluded also that lymphoid tissue is the most sensitive to benzene. Irons et al.[110] showed that administration of benzene to rats, mice, and rabbits produces a dose-dependent decrease in the number of lymphocytes which may precede the decrease of the other blood cells.

As it is known, lymphocytes are the immunopotent cells of the body. On the other hand, lymphocytes are responsible for mediation of both the humoral and cellular immune responses. Lymphocytes are divided into two major types: (1) B lymphocytes, responsible for mediating humoral immune response and/or antibody production; and (2) T lymphocytes, responsible for cell-mediated immune responses. Irons et al.[110] showed that benzene administration, in vivo in mice, which affects lymphocyte function is observed at doses of the compound that produce little or no measurable differences in the number of circulating cells. Irons et al. administered 44/mg/kg of benzene to mice for 3 days. In animals, plaque-forming cells which measure the number of cultured lymphocytes, producing IgM antibody, were considerably reduced at 132 mg/kg benzene whereas spleen cellularity significantly decreased at high doses. According to Irons et al., these results are contrasted by the changes in the number of circulating lymphocytes which were decreased only at doses of 440 mg/kg or higher. Furthermore, lymphoproliferative responses (LPR) of mouse spleen B lymphocytes, stimulated in culture with lipopolysaccharide derived from E. coli (LPS), were 40% of control in animals treated 3 days at the 264-mg/kg dose. Similarly, LPR of spleen T lymphocytes, stimulated with concanavalin A which causes T-lymphocyte proliferative response, were 38% of control spleen cells in mice administered 264 mg/kg benzene. Irons et al.[110] made the same trials with phenol, hydroquinone, and catechol. Although phenol produced no evidence of immunotoxicity, both hydroquinone and catechol were found to be immunosuppressive in mice when administered either intravenously or intraperitoneally.

On the other hand, the patterns of immunosuppression exhibited by hydroquinone and catechol were significantly different. Hydroquinone reduced spleen and particularly bone marrow cellularity. Contrary to this, catechol reduced spleen cellularity, but it had no appreciable effect on cellularity of bone marrow. Hydroquinone suppressed both LPS and LPS plus dextran sulfate LPR in bone marrow and produced a corresponding decrease in LPS and LPS plus dextran sulfate-stimulated plaque-forming cell assays. Catechol showed very little effect on bone marrow LPR or LPS, and dextran sulfate stimulated plaque-forming assay. Despite this, it produced a marked suppression of LPS-stimulated plaque-forming cell assays. According to Irons et al.,[110] these results indicate that hydroquinone affects LPR, spleen and bone marrow cellularity, and bone marrow progenitors to antibody-producing cells. Catechol appears to have little effect on LPR, but interferes with endstage of plaque-forming cell assay. In other words, catechol has a selective effect in vivo on endstage effect or cell differentiation in the absence of a generalized effect on lymphocyte proliferation.[110,111]

On the other hand, Moszczynski and Lisiewicz[112] determined T lymphocyte functions in 72 workers exposed to benzene, xylene, and toluene of comparatively low levels during 31 and 122 months. The workers, after a service of 55 to 122 months, exhibited a decrease in T lymphocyte count without, however, changes in their functions. The investigators emphasized the importance of the E rosette test in the evaluation of early toxic effects of benzene and its homologs on the lymphocyte system. Furthermore, they postulated that the reduction of T lymphocytes might be due to the depressive effect of benzene, especially of its free radical form, benzene-epoxide, on the lymphocyte system. Furthermore, Moszcyzynski and Lisiewicz,[113] in another study on 106 workers exposed to benzene and its homologs, reported that the lowered count of lymphocytes was due to decreased numbers of T, non-T, and non-B cells.

Yet, in another study, Moszczynski and Lisiewitz[114] showed that prolonged exposure to benzene and its homologs results in decreased numbers of lymphocytes having intact N-acetyl-beta-D-glucosaminidase (AG)-positive lysosomes.

2. Lymphocytosis

Bernard[115] described three cases of chronic benzene toxicity associated with severe lymphocytosis, and three cases with moderately increased lymphocyte counts. The lymphocyte ranged in severe cases between 15 to 27 × 10^9/1, the relative lymphocyte values being in the range of 82 to 87%. The lymphocytes decreased a few weeks after the patients were removed from exposure to benzene. No other reasons could be found for the lymphocytosis observed in these patients. Results of marrow puncture performed on one of them showed no alterations. The present author and associates have found mild lymphocytosis in two workers with chronic benzene toxicity.[64] In one of them, lymphocytosis was the only abnormality detected.

3. Decrease in the Leukocyte Alkaline Phosphatase Activity in Individuals with Chronic Benzene Toxicity

Girard et al.[116] investigated the leukocyte alkaline phosphatase activity (LAP) in 319 workers who were chronically exposed to moderate or low levels of benzene, such as 10 and 25 ppm, respectively, and in 204 controls. During 3 years, a total of 827 tests were performed. In a great part of the individuals who were exposed to benzene or toluene, the LAP activity was found to be reduced. Girard et al. suggested that benzene and toluene, similar to some hematologic disorders such as chronic myeloid leukemia, have an effect on LAP activity. In addition, according to Girard et al., the estimation of LAP activity is a reliable group test in the surveillance of exposed subjects.

Contrary to the study of Girard et al., Songnian et al.,[117] in China, obtained an increase in the LAP activity in 175 workers exposed to a concentration of benzene of about 100 mg/m³. Songnian et al. have performed experiments with rats exposed for 6 months to high levels of benzene. Alkaline phosphatase activity in the bone marrow of benzene-intoxicated rats was also increased. According to Songnian et al., LAP assays are more sensitive than leukocyte counts in the diagnosis of chronic benzene toxicity. This problem deserves further investigations.

4. Changes in the Leukocyte Osmotic Resistance in Individuals Exposed to Benzene

Pollini and Colombi[118] showed a decrease in the leukocyte osmotic fragility in 14 individuals with chronic benzene toxicity. Interestingly, according to Pollini and Colombi, the changes in the leukocyte osmotic fragility are present before the appearance of hematological abnormalities due to chronic exposure to benzene.

5. Other Abnormalities in Leukocyte Function and Enzymes of Leukocytes

Doskin[63] established a decreased phagocytic function of neutrophiles in 152 workers with chronic benzene toxicity. On the other hand, Volkova[119] determined a reduced phagocytic activity of rabbit leukocytes by the daily 3-hr inhalation of 0.02 mg/ℓ of benzene (6 ppm). This effect became apparent before other signs of chronic benzene toxicity were evident. Koslova and Volkova[120] determined the phagocytic activity of leukocytes in 252 workers in a leather factory. According to Koslova and Volkova, a definite relationship has been noted between the blood picture and the benzene content in the air of the workshop. When the concentration of benzene in the air decreased, the blood picture became more normal. The improvement however, largely depends both upon the previous length of exposure to benzene and the blood changes. Furthermore, Koslova and Volkova considered the phagocytic activity of leukocytes as a very sensitive test for the effect of benzene in humans. They also noted that where the concentration of benzene in the air of the workplace surpassed the permissible level by 1.5 to 2.5 times, in 80% of the workers a drop in the phagocytic activity of leukocytes could be noted.

On the other hand, Boiko and Makarieva[121] demonstrated reduced glucogen content and inhibited activity of peroxidase in the neutrophils of the workers engaged in the production of benzene and its homologs from crude oil in the U.S.S.R. Furthermore, Moszczynski[122] has performed a study on the enzymatic content of neutrophils of 66 workers who were in contact with benzene and its homologs for 20 to 122 months. The following cytochemical parameters were studied in neutrophils: alkaline and acid phosphatase, beta-glucoronidase, N-acetyl-glucosaminedase, myeloperoxidase, lactic dehydrogenase, and the glycogen and lipid contents. In all workers exposed to organic solvents, an increase of the acid phosphatase and beta-glucoronidase activity, as well as a decrease of the glucogen content, were noted whereas the decrease of the alkaline phosphatase and myeloperoxidase activity, as well as that of the lipid content, was noted only in the workers exposed to higher concentrations of benzene and its homologs. Yet, there was no correlation between the neutrophil enzymatic content and duration of exposure to the solvents. According to Moszczynski, it seems that PAS reaction has a special practical significance in addition to the estimation of T cell content in the peripheral blood.[122]

On the other hand, Moszczynski and Lisiewicz[123] studied 39 workers from the same workplace mentioned above. The lymphocyte-associated immunity was examined by using such parameters as total lymphocyte count, T and B cell count, blastic transformation test, tuberculin test, immunoglobulin concentrations, etc. The only abnormal findings were a reduced number of total lymphocytes, those of the Ea and the E_{18}. The rosettes (T cells) compared to a control group of 38 subjects working in the same industrial establishment but not exposed to any chemicals.

Furthermore, Lange et al.[124] determined the leukocyte agglutinins in 35 persons exposed to benzene. The leukocyte agglutinins were demonstrated in the sera of 10 workers. The increase of leukocyte agglutinine per liter of sera after their incubation of benzene and its homologs was also demonstrated.

O. Eosinophilia

Some investigators have reported eosinophilia in chronic benzene toxicity. Heim de Balzac and Lafont[125] reported that eosinophilia may occur without concomitant changes in other blood parameters. Duvoir and Derobert[126] have corroborated this. In their study, relative eosinophil values greater than 5% were found in 21.8% of the 555 workers. One case had 18.1% eosinophils. Savilahti[40] found eosinophils above 6% in 18 workers. In one case the eosinophils reached 15.5%. Hernberg et al.[127] studied the eosinophil values in 120 out of 144 workers who were exposed heavily to benzene 10 years prior.[40] According to this study, following a 10-year cease of exposure, the eosinophilia had evidently decreased, thus suggesting that the high rate of eosinophilia, ranging between 0 and 15.5%, was due to the effect of benzene. Sungur[128] reported eosinophils above 3% in 28.2% of the workers studied. It should be pointed out that all these studies accept low eosinophil values as normal. Today, relative eosinophil values up to 8% have come to be considered normal.[129] In our study we found eosinophilia in 5 workers, that is 2.3% of the 217 workers.[64] Eosinophil values were in the range of 8 to 39%. If the one case involving both basophilia and eosinophilia is added to this number, the ratio becomes 2.76%. However, this finding was not statistically significant.

Goldwater, Smith, and Browning did not find eosinophilia in their studies.[102,130,19] Hamilton-Paterson and Browning[103] measured eosinophils in 200 individuals who were working with benzene and in 200 control individuals using the direct method and found no difference; all values were in the normal range. In conclusion, we would like to say that eosinophilia is a rarely encountered finding in chronic benzene toxicity. Therefore it is an unreliable indicator. We reach this conclusion from the material available in the literature and from our own studies.

P. Basophilia

Smith[130] reported that basophilia is seen in conjunction with chronic benzene toxicity. Yet the other investigators did not confirm this finding.[19,64] Smith assumed the normal range to be one basophil per 200 to 300 leukocytes and found that this ratio increased to 1 to 2% in benzene toxicity cases. However, today the accepted normal ratio of basophils is at least 0 to 2%.[129]

The present author and associates found that the fraction of basophils varied between 0 to 2% in the 100 controls tested.[64] Only two of the 217 workers, that is 0.92%, had basophils above 2%. There was no significant differences between the values of the normal control subjects and those of the workers under investigation.[64] In our laboratory, a study using the direct counting method of absolute numbers of basophils was performed in eight pancytopenic workers with chronic benzene toxicity.[131] The absolute numbers of basophils in normal were between $30.5 \pm 14.20/m^3$. In six patients with bone marrow depression due to chronic benzene toxicity, absolute basophil numbers decreased to 8.16 ± 6.30 ($p < 0.01$). Yet there was no significant difference in the absolute basophil numbers (counting) of normals and 8 cases with chronic benzene toxicity without bone marrow depression.[131]

In conclusion, basophilia is a rare occurrence in chronic benzene toxicity and has no statistical significance in its diagnosis. Despite this, we have an impression that the occurrence of basophilia is a possible sign on the presence of a hematologic malignancy or that of a blood dyscrasia due to chemicals including benzene.[132]

Q. Monocytosis

Monocytosis is not considered to be an indicator of chronic benzene toxicity.[19,40,102,103]

This author and associates found only one case which exhibited monocytes more than 8%.[64] The accepted normal range of monocytes is between 2 to 8%.[129] Sungur[128] found monocyte values greater than 10% in 12.1% of the workers studied. On the other hand, Roth et al.,[133] in Romania, performed a hematologic study in 850 workers with benzene exposure. An absolute monocytosis was found in 5.4 to 22.6% of them. Additionally, in their study, Roth et al. emphasized that some monocytes resembled monoblasts or had the appearance of immature monocytes. According to the investigators, the changes in lymphocytes and monocytes may be considered as the early signs of benzene toxicity and it can be used in the early detection of toxicity. Furthermore, in one of our cases of paroxysmal nocturnal hemoglobinuria due to chronic benzene exposure, the present author and associates observed a striking monocytosis during the course of his illness.[134] This patient showed a monocytosis reaching 24%. It persisted in the course of the disease and was present until the end of its fatal course.

The findings of bone marrow were not consistent with the diagnosis of monocytic leukemia. These facts would indicate that there may be some benefit in reevaluating the relationship between monocytosis and chronic benzene toxicity, but it should also be remembered that young lymphocytes, especially in response to irritation, can resemble monocytes. We frequently encountered such lymphocytes in the peripheral blood of the workers that were exposed to benzene; we called them young or atypical lymphocytes.[64] On the other hand, in their study among 100 workers exposed to benzene and its homologs, Moszczynski and Lisiewicz[113] have established an increased count of monocytes compared to controls.

R. Thrombocytopenia

Reports on the decrease of the platelets in response to chronic benzene exposure have been variable. According to Savilahti[40] it is the most frequently encountered hematological change. He found platelet counts less than 200,000 in 62% of the workers. Goldwater[102] obtained similar results; 32% of his cases had thrombocytopenia. It was the third important change he found after anemia and macrocythemia. In this study, thrombocytopenia was defined as a platelet count under 100,000. Similarly, according to Doskin,[63] the hematological disturbances in the majority of cases of chronic benzene toxicity commenced with thrombocytopenia. Platelets ranged in their thrombocytopenic workers between 96 and 155 $\times$ 10^9/l. On the other hand, Grazioli and Monteverdi[135] were unable to detect thrombocytopenia in most of their mild cases of chronic benzene toxicity. Sungur[128] found platelet counts less than 150,000 in 7.1% of the workers studied. In the study of the present author and associates on 217 workers with chronic benzene toxicity, the total ratio of cases involving thrombocytopenia was as 9.21%. Of these, 1.85% had thrombocytopenia alone, 4.6% had thrombocytopenia accompanied with leukopenia and 2.76% had thrombocytopenia as a part of pancytopenia.[64] In this study, the percentage of thrombocytopenia was less than that of leukopenia.

A recent study performed among 231 workers exposed chronically to benzene in 40 small or large workshops in Istanbul and Izmit showed thrombocytopenia was present in 2.2% of the workers.[132] Furthermore, leukopenia was found in 3.9% of the workers, and pancytopenia was present in only 0.54% of the individuals. The percentage of benzene was more than 1% in 76.4% of the solvents, and 19.1% of the thinners which were used in the workplaces studied. As we shall discuss in the section on severe benzene toxicity in further detail, the present author and associates found that 4 out of 44 benzene-mediated aplastic anemia cases had platelet counts within normal limits while exhibiting moderate or severe anemia and leukopenia.[37,87] We believe that benzene has a greater effect on leukocyte production than on the production of megakaryocytes, i.e., platelets. Grazioli and Monteverdi[135] reported that chronic exposure to benzene affects coagulation by damaging platelets, without causing thrombocytopenia. This author and associates measured the adhesiveness, agglutinability, and viscous metamorphosis of platelets in some of the cases during the survey on 217 workers.[64] We were unable to detect any characteristic differences. The only abnormality relating to platelets was the occurrence of giant platelets in one worker exposed chronically to benzene.

In literature, there are few investigations concerning the presence of the decreased platelet functions in chronic benzene toxicity.[136-138] Craveri[137] studied prothrombin time, the fibrinogen level, fibrinolytic activity, and thromboelastogram in 18 individuals associated with mild or moderate hematological findings of chronic benzene toxicity. This investigation showed inconstant findings with respect to these parameters. According to Craveri, the hemorrhagic effects of chronic benzene toxicity are not solely due to thrombocytopenia, but also to increased fibrinolytic activity. Furthermore, Saita et al.[138] have performed thromboelastographic studies in eight patients associated with varying degrees of thrombocytopenia due to chronic benzene toxicity. According to Saita et al., there may be a platelet abnormality in thrombocytopenia influencing the agglutinating activity of platelets rather than their thromboplastinic action and their effect on coagulation time.

S. Erythrocytes and Hemoglobin

Contradictory results have been reported on the early effect of benzene on erythrocytes. According to Goldwater,[102] chronic benzene toxicity affects the erythrocytes first, causing a decrease in the erythrocyte number. A decrease in hemoglobin then follows, which leads to macrocythemia. Goldwater found decreased erythrocytic counts in 47.9% of the workers tested. Savilahti[40] found anemia in 35% of the workers. Browning[19] reported that anemia does not occur in the early stages of chronic benzene toxicity, but it is seen only in cases that have progressed. Browning also stated that a mild and easily treated anemia may be associated with the early phases of chronic benzene toxicity. The type of anemia involved is still an area of controversy. Goldwater[102] claims that it is macrocytic anemia. Savilahti[40] thinks that it is hyperchromic anemia.

On the other hand, Kuhbeck and Lachnit[139] studied the problem of macrocytosis in workers exposed to benzene in a tire factory. They measured mean corpuscular volume (MCV) of erythrocytes, but were unable to show significant macrocytosis in the exposed group, as compared to the control group. The investigators believed that macrocytosis does not necessarily have to accompany chronic benzene toxicity. This author and associates found that 33.1% of the 217 workers with chronic benzene exposure had hemoglobin values below 12 g and 32.7% having hematocrit under 40%.[64] The MCV of the workers that exhibited mild or moderate anemia was between 86 and 96 fℓ. This value was never greater than 100 fℓ. This would imply that we cannot find macrocytosis in the early phases of chronic benzene toxicity. It is reported that 31% of the workers exhibiting mild or moderate cases of anemia, responded to a brief, orally administered iron therapy.[64] Their hemoglobin rose rapidly, reaching normal values after the administration of iron. Although this result has led researchers to conclude that anemia was mainly due to iron deficiency, we would like to point out that only 17% of the individuals tested as controls had less than 12 g hemoglobin, and only 17% had hematocrit less than 40%. The total picture would imply that individuals exposed to benzene are more anemic and that benzene mediates this anemia.[64]

According to Doskin,[63] disturbed erythropoiesis was the first change in some cases. Changes in erythrocytes were characterized by normochromic anemia, with a tendency toward hyperchromia 1 year after the beginning of employment. Later, when the benzene concentrations were reduced to the maximum permissible concentration, the picture underwent a change towards normal and even high red blood cell levels. Furthermore, Moszczynski and Lisiewicz[113] have noted that according to the results of their study on 106 workers occupationally exposed to benzene, relatively low doses of benzene and its homologs caused a decrease in the mean corpuscular hemoglobin concentration, possibly as a result of a disturbance of hemoglobin synthesis. Contrary to these, some investigators have suggested that polycythemia, rather than anemia, occurs in mild benzene toxicity. Hunter[140] has reported that an increase in erythrocyte counts were found in some cases.

T. Bone Marrow in Individuals with Chronic Benzene Toxicity

The state of bone marrow, particularly in the patients with aplastic anemia due to chronic benzene exposure, will be discussed in detail later. Here we shall only summarize our results in 11 individuals with chronic benzene toxicity.[64] The state of the bone marrow is investigated in 11 workers, four with leukopenia associated with thrombocytopenia, one with lymphocytosis, one with acquired pseudo-Pelger Huet anomaly, and five with pancytopenia which were diagnosed during the survey among apparently normal-appearing active workers.[64] The results are summarized in Table 3. No remarkable changes were observed in two of them. In only one of the remainder, bone marrow was hypercellular with maturation arrest in the erythroid and myeloid series. In another case, the bone marrow was slightly hypercellular with the arrest of maturation in the granulocytic series. In two workers, the bone marrow was normocellular, and in two, slightly hypocellular with maturation arrest in either erythroid or myeloid series. In another three, the bone marrow showed marked vacuolization as observed in some cases of chloramphenicol-induced pancytopenia.[129]

A great variation in the bone marrow findings has been reported both in animals poisoned with benzene experimentally, and in human beings with signs and symptoms of benzene toxicity. The same variation was detected to a lesser extent in the 11 workers in our series in whom features accompanying the early stages of chronic benzene toxicity were present. In the majority of our cases, the bone marrow smears were normocellular; in the others, they were either hypocellular or hypercellular. Some of them also had a maturation arrest either in the erythroid or myeloid series or in both.[64] In some, the only abnormal finding was the vacuolization detected in the myelocytes or plasma cells (case 6). Furthermore, in two subjects there were no changes in the bone marrow smears, in spite of leukopenia associated with thrombocytopenia or pancytopenia (cases 1 and 8).

U. Changes in Iron Metabolism in Chronic Benzene Toxicity

Observations in both animals and humans indicate a change in iron metabolism in the cases of chronic benzene toxicity. Villiani and co-workers[141] studied this problem in animals using radioactive isotopes, and found a decrease in the iron utilization of hematopoietic tissue. Chronic benzene toxicity causes no changes in iron uptake by reticulocytes and does not alter the daily renewal of erythrocytes. On the other hand, benzene inhibits iron utilization of basophilic erythroblasts.

Truhaut and associates[142] have obtained inconstant results in animal experiments with ^{59}Fe. In some animals, the rate of iron metabolism increased, but the rate of binding by erythrocytes was not abnormal. This resulted in imperfect erythrocytes that were broken down in periphery. Snyder and Kocsis[143] studied the effect of benzene on erythropoiesis by using the incorporation of ^{59}Fe into hemoglobin of maturing red blood cells as a measure of erythrocytes synthesis in mice. They showed that after a single dose of benzene, a reduction in the incorporation of ^{59}Fe into red cells occurred when leukocyte production was normal according to cell-counting techniques. This problem is discussed in Chapter 4.

V. Blood Sulfhydryl Group Concentration and Chronic Benzene Toxicity

Paolino and co-workers[144] induced acute benzene poisoning in mice and demonstrated that blood sulfhydryl concentrations decreased. The renal sulfhydryl concentrations decreased while hepatic sulfhydryl increased. These changes were particularly evident in the initial stage of poisoning. The decrease in sulfhydryl concentration results in the utilization of organic sulfur compounds. These findings indicate that the sulfhydryl groups distributed in the organism enable sulfoconjugation upon exposure to a toxin.

Table 3
RESULTS OF BONE MARROW IN 11 APPARENTLY HEALTHY WORKERS USING MATERIALS CONTAINING BENZENE

Case	Type of hematological changes	Bone marrow findings			
		Cellularity	Myeloid series	Erythroid series	Megakaryocytes
1	Leukopenia and thrombocytopenia	Normocellular	No change	No change	Present
2	Leukopenia and thrombocytopenia	Normocellular	Maturation arrest	Vacuolization	Decreased in number
3	Leukopenia and thrombocytopenia	Slightly hypocellular	No change	No change	Present
4	Leukopenia and thrombocytopenia	Slightly hypercellular	Maturation arrest	No change	Present
5	Pseudo-Pelger-Huet anomaly	Hypercellular	Marked maturation arrest	Slight maturation arrest	Present
6	Lymphocytosis	Normocellular	Vacuolization in myeloid series	No change	Present
7	Pancytopenia[a]	Normocellular	Maturation arrest	Maturation arrest	Present
8	Pancytopenia[a]	Normocellular	No change	No change	Present
9	Pancytopenia[a]	Normocellular	Maturation arrest vacuolization in myelocytes and plasma cells	No change	Present
10	Pancytopenia[a]	Slightly hypocellular	Maturation arrest vacuolization in myelocytes and plasma cells	No change	Present
11	Pancytopenia[a]	Slightly hypocellular with mild increase in lymphocytes	No change	No change	Present

[a]   These five pancytopenic patients were detected during this study. These pancytopenic individuals were active in their jobs when we detected them. The interesting point was that pancytopenia had not progressed sufficiently to give signs and symptoms of aplastic anemia.

From Aksoy, M., Dinçol, K., Akgün, T., Erdem, Ş., and Dinçol, G., *Br. J. Ind. Med.*, 28, 296, 1971. With permission.

W. The Effect of Benzene on Porphyrine Metabolism

Biancacchio and Fermariello[145] studied porphyrine metabolism in rats poisoned with benzene. They showed that these circumstances led to a 6-fold increase in protoporphyrine and a 2.5-fold increase in urinary coproporphyrine levels. They concluded that the changes in subacute benzene toxicity in animals relate to protoporphyrine. On the other hand, Kahn and Muzyka[146] have administered benzene at 0.1 mℓ/kg of weight, 4 times a week for 5 to 6 months to rabbits. Despite the normal hematological data, a considerable increase in the delta-levulinic acid (ALA) of erythrocytes was observed. At the same time, an increase in the content of ALA, coproporphyrine, and porphobilinogen in the grey matter of the brain were found. The investigators studied 121 workers exposed to 5 mg/m^3 of benzene and toluene.[146] In 74 workers, there were complaints indicating functional disarrangement with the activities of the central nervous system such as headache towards the end of the shift, rapid exhaustion, and disarrangement in sleep, etc. Also, 60% of the workers had increased ALA in the erythrocytes and 33% had increased coproporphyrine in urine. There was no correlation between the occurrence of abnormalities relating to porphyrine metabolism and the duration of exposure to benzene. According to Kahn and Muzyka, a good correlation exists between the disorders of porphyrine metabolism and functional disarrangement of the central nervous system. Kahn and Muzyka emphasized that the disorders of porphyrine metabolism due to chronic benzene toxicity take place before the appearance of hematological abnormalities.

X. Changes in Erythrocyte Enzymes and Chronic Benzene Toxicity

Zatonski[147] investigated erythrocyte metabolism in 99 individuals exposed to benzene. According to the investigator, there was a small, but statistically significant decrease of glucose-6-phosphate (G-6-PD) activity in these individuals. In 60% of these individuals with benzene exposure, the G-6-PD activity was lower than normal values or equal to the lower values within normal limits. The greatest deviation has been found for the activity of phosphofructokinase which is the key enzyme of anaerobic glycolysis. According to Zatonski, the decrease of the activity of erythrocyte phosphofructokinase should be considered as one of the important factors in the development of increased hemolysis found in some cases of chronic benzene toxicity.

Y. Changes in Some Enzymes in Chronic Benzene Toxicity

The activity of serum alkaline phosphatase has been investigated by Hancke[148] in a group of 35 workers, who were exposed to a concentration of benzene higher than the permissible level, half a year after the discontinuation of exposure. In 18 out of 35 workers, there was a drop in alkaline phosphatase activity showing the lowest limit of normal range. The difference found between the mean value for the control group and that for the group under the test was statistically significant. Hancke also studied the levels of the serum alkaline phosphatase, aldose, and glutamic oxalic transaminase in 7 individuals with hematological abnormalities due to chronic benzene exposure. A drop in serum alkaline phosphatase activity and, in some of them, also a decrease in serum glutamic oxalic transaminase were found.

Z. Changes in Serum Immunoglobulin and Serum Complement Levels in Chronic Benzene Toxicity

Lange et al.[149] studied immunoglobulin levels in workers exposed to benzene, toluene, and xylene of varying degree. They found that serum IgG and IgA levels were decreased. Contrary to this, IgM levels were found to be in normal range or slightly elevated. Similarly, Smolik et al.[150] determined serum complement levels in 79 persons exposed occupationally to benzene, toluene, and xylene. In 62 of the workers, complement level was lower than the mean value of the control group. According to the investigators, these findings, with the changes in immunoglobulin levels, suggest the involvement of immunologic factors in the pathologic mechanism of exposure to benzene and its homologs.

AA. Changes in Bioelectric Activity of the Brain in Chronic Benzene Toxicity

Sobczyk et al.[151] performed electromyographical (EEG) recordings in 100 female workers who were exposed to solvents containing benzene. Mean age of these individuals was 41.2 years. EEG recordings were obtained 3 to 4 years following hematological changes and increased phenol excretion. In 40 out of 100 workers, EEG examinations showed abnormal findings such as changes in temporal region (50%), paroxysmal bursts (17.5%), and generalized changes (32.5%). Sobczyk et al. concluded that changes in bioelectric activity of the brain may be of value as an index of chronic benzene toxicity.

BB. Skin and Chronic Benzene Toxicity

Although an overwhelming part of benzene is absorbed by inhalation, a small or a possibly insignificant part can be taken through the skin. Dermal exposure is only a minor source of concern since benzene is poorly absorbed by skin.[152] Percutaneous absorption of benzene may happen via direct skin contact with benzene solution or solvents containing benzene. According to Hanke et al.,[153] 0.4 mg/cm² of benzene was absorbed per hour after applying liquid benzene to skin. Maibach[154] has recently reported the results of painting radiolabeled ¹⁴C-benzene on the forearm and palm of human volunteers. Penetration of benzene ranged from an average of 0.06% of the dose when applied to the forearm and that of 0.13% when applied to the hand. Abraded skin or prolonged contact with skin would allow the penetration of a relatively greater amount of benzene. Similarly, Wester and Maibach[155] applied 3.5 mg ¹⁴C labeled benzene per square milliliter to forearm of rhesus monkey and showed that only 0.2% (7 mg/mℓ²) was absorbed percutaneously. On the other hand, when the skin is injured, the absorption of this chemical is accelerated.[156] According to Täuber,[157] if solutions of benzene in other organic solvents are applied to the skin, the total absorbed component is even lower. But this conclusion needs experimental evidence. Furthermore, there is no convincing evidence today on the absorption of benzene vapor from the air through the skin.[157]

Recently, Suston et al.[158] conducted at the request of OSHA, a series of in vivo studies by applying undiluted benzene or rubber solvent containing 0.5% benzene to the hairless mice. The results of these studies showed that benzene is absorbed to the extent of approximately 1% of the applied dose which was available for absorption. Suston et al. estimated that under job conditions observed during tire building, a worker dermally exposed 150 times per day to rubber solvent of 0.5% benzene will absorb approximately 6 mg of benzene. This represents a significant addition to the 14 mg of benzene which is estimated to be retained in the body following inhalation of benzene at a concentration of 1 ppm. The fact that the skin on the hands of tire workers is often cracked and fissured almost assures that the penetration of benzene contacting the skin will be rapid and greater than would be expected for intact skin. According to Suston et al.,[158] whether or not these results would represent significant increased risk to health remains to be determined.

III. APLASTIC ANEMIA DUE TO CHRONIC BENZENE TOXICITY

The total bone marrow depression that occurs in advanced cases of chronic benzene toxicity will exhibit itself as pancytopenia in the peripheral blood, reflecting mostly the picture of classic aplastic anemia. As we shall see later, aplastic anemia that occurs in response to chronic benzene toxicity will differ in some respects from the clinical and hematological picture of classic aplastic anemia. Furthermore, it can exhibit a variety of clinical and hematological pictures. The onset of the disorder is very difficult to detect. The patients usually do not seek medical help until it is too late, that is, about the time a total bone marrow depression occurs. This author and associates found 5 out of 46 patients with aplastic anemia during the screening of workers whose jobs involved contact with benzene.[37,64,87]

The elusive progress of the disorder is not the only reason for the delay in seeking medical help. In our studies we observed that individuals that were chronically exposed to benzene were also unconcerned with their health. In other words, they were aloof to the disorders they may have noted.

A. Frequency

According to some studies, the frequency of aplastic anemia, either idiopathic or induced by chemicals or drugs is 2 to $5/10^6$ persons per year in the Western world.[159,160] This incidence was reported to be as high as $12/10^6$ persons per year in some other studies.[43,159-163]

Recently, Heimpel[163] reported more lower frequency data for some cities of Europe including Israel. He pointed out that, according to the joint study of the International Agranulocytosis and Aplastic Anemia Group, the frequency of aplastic anemia in Western countries changes between 1.4 and 3.1×10^6/year.[163] The lowest data in this study was obtained in Israel. On the other hand, it seems probable that the prevalence of aplastic anemia is higher in Eastern countries. According to Böttiger,[43] aplastic anemia is at least four or five times more common in the East than in the West. On the other hand, Böttiger[43] has considered that the difference in the prevalence of aplastic anemia between West and East has a multifactorial explanation. In this respect, he accuses, except the possible role of genetic differences in susceptibility, other exogenous factors such as occupational hazards including benzene, high incidence of hepatitis, and exposure to ionizing radiation, etc.[43]

Following the control on the use of benzene in workplaces in the Western countries, the annual number of aplastic anemia due to this chemical decreased rapidly. In other words, in recent years benzene, like chloramphenicol, has been displaced from the top of the list of aplasia-inducing agents in many countries of the Western world.[43] Contrary to this, the role of benzene in the etiology of aplastic anemia in Eastern countries including Japan is still important.[43] For example, according to some Japanese studies, organic solvents were suspected as causative factors in 6 to 12% of aplastic anemias.[43] Similarly, Whang[164] found a history of benzene exposure in 8.6% of the patients of his Korean material. In contrast to data obtained in Western countries, in Turkey the most important causal agent for the development of aplastic anemia in a series of 108 patients with this hematologic disorder was benzene, which accounted for 25 (23.1%) of the cases.[47] Although "epidemics of chronic benzene intoxication" in this country disappeared following the prohibition of benzene in 1969, the persistence of aplastic anemia due to benzene in Turkey is either due the use of this aromatic hydrocarbon as a solvent for rubber, dyes, or for cleaning in some workplaces, or to exposure to this chemical agent as a solvent before its use was regulated.[47] Recently, in a study on the benzene contents of solvents and thinners, etc. used in the workplaces in Istanbul and its vicinities, Aksoy et al.[132] established that the 75.4% of the solvents and 19.1% of the thinners used contained benzene of more than 1%, ranging between 1.1% and 7.6% (mean: 2.88%).

B. Clinical Findings

The first findings are quite variable. They usually depend on the severity of the pancytopenia. In some cases, the initial clinical symptoms will be similar to the symptoms of anemia. They are exhaustion, shortness of breath, palpitation, palor, and sometimes insomnia. In some cases, the individual seeks medical help because of fever, usually resulting from secondary infections that occur because of leukopenia. In some cases the individual will see a physician because of thrombocytopenia and the various manifestations of hemorrhagic diathesis. It causes such manifestations as bleeding from the nose, gums, skin (purpuric spots and echymoses), bleeding in the digestive, urogenital, and gynocological or even neurological systems. In mild or early cases, if exposure to benzene is eliminated and correct treatment initiated, a period of slow recovery will begin. In contrast, all symptoms associated with aplastic anemia will appear in severe cases. These are primarily a moderate or severe anemia, secondary infections that may lead to sepsis, and hemorrhage from every region of the body.

C. Anemia

Aplastic anemia that is caused by chronic benzene toxicity is similar to other types of aplastic anemia, idiopathic or caused by various agents. As we shall see below, some cases of benzene-mediated aplastic anemia will, in addition, manifest the findings of increased hemolysis. The anemia is usually normochromic and normocytic. In some cases it can be hyperchromic macrocytic. If serious hemorrhage has taken place, it can be very slightly hypochromic. In certain cases, the mean corpuscular volume (MCV) reaches 115 fℓ, indicating megaloblastic bone marrow. In 4 out of 44 patients with aplastic anemia due to chronic benzene exposure, MCV ranged between 103 and 118 fℓ and there were megaloblastic bone marrow in these patients.[37,87] Megaloblasts varied between 4% and 22% in their bone marrow smear.[37,87] Occasionally, anemia may be very serious. In our series comprising 44 individuals with benzene-induced aplastic anemia, red blood cells, hemoglobin, and PCV ranged between 0.64 and 3.6 $\times$ 10^{12}/ℓ; 2.4 and 11.2 g/dℓ; and 0.055 and 0.36, respectively. The morphological changes of the erythrocytes are not usually extensive. Ovalocytosis, macrocytosis, few hypochromic erythrocytes, few target cells, and rarely, poikylocytes may be found. Nucleated red blood cells may be present in the peripheral blood film of some cases. In six patients of our series, nucleated red blood cells were present in the peripheral blood ranging between 1 and 8/100 WBC.

The present author and associates had the impression that this phenomenon, namely erythroblasthemia, is more frequent in benzene-mediated aplastic anemia than in the other types of aplastic anemia. The causes of the erythroblastosis can be attributed to various mechanisms such as ineffective erythropoiesis, folic acid deficiency, and increased hemolysis.[87] One of our six patients with benzene-induced aplastic anemia showed a striking erythroblasthemia. Nucleated red blood cells in peripheral blood smears ranged in the course of his illness between 8 and 32%. The cause of this phenomenon was possibly folic acid deficiency since the nucleated red blood cells in the peripheral blood smear disappeared following folic acid therapy. Additionally, in another pancytopenic patient with chronic exposure to benzene, the concentration of folic acid was found to be low.[87] In 15 out of 44 patients with benzene-induced pancytopenia, anemia was macrocytic, in 26, normocytic, and in three, microcytic. In three of them, the cause of hypochromic microcytic anemia was the coexistence of heterozygous beta-thalassemia.[37,50,87] If a thalassemic individual develops aplastic anemia with chronic benzene toxicity, then beta-thalassemia associated changes in erythrocytes and hemoglobin profile can still be detected, like hypochromia, microcythemia, and numerous target cells, and increase in hemoglobin A$_2$, etc. In two pancytopenic patients, one fatal, the level of fatal hemoglobin was very high.[50] This problem will be discussed later.

On the other hand, according to some investigators, macrocytosis is the most frequently reported erythrocyte abnormality following exposure to benzene.[102,165-168] The cause of macrocytosis is not known. A block in nucleoprotein metabolism, in other words, an effect on DNA synthesis or folic acid deficiency, may be responsible for some benzene-induced bone marrow depression associated with macrocytosis.[87,102,165-168] According to Gorini et al.,[166] the altered DNA synthesis may cause a prolongation of mitotic interphase period. This phenomenon causes the production of macroerythroblasts which results in macrocytes. The occurrence of macroerythroblasts has been noted in the bone marrow by Gorini et al.

and Curletto and Ciconali.[166,167] According to Corsico et al.,[168] macrocytosis of erythrocytes is of prognostic importance. On the other hand, according to Helmer,[169] an advanced anemia without combination of thrombocytopenia is not due to chronic benzene toxicity. This conclusion is not correct because sometimes the absence of thrombocytopenia or leukopenia with the presence of anemia due to chronic benzene toxicity might be encountered.

D. Manifestations of Increased Hemolysis in Chronic Benzene Toxicity

As it is known, the number of the reticulocytes in aplastic anemia is either in normal range or decreased. Following the improvement of bone marrow depression, the number of reticulocytes starts to increase. Contrary to this, the presence of a mild reticulocytosis and a mild bilirubinemia was first reported by Goldwater[102] in cases of chronic benzene toxicity. Later, Erf and Rhodas[165] added increased urinary and fecal urobilinogen to this list. In out of 32 patients with benzene-mediated aplastic anemia, we found a mild reticulocytosis in 16 patients.[87] The reticulocytes ranged between 2.1 and 4.8% in these cases. Recently, Moszczynski and Lisiewicz[113] reported an increased number of reticulocytes in 108 workers occupationally exposed to low or moderate levels of benzene and its homologs compared to controls.

The occurrence of reticulocytosis in chronic benzene toxicity may be explained either by ineffective erythropoiesis present in benzene-mediated aplastic anemia or by the occurrence of increased hemolysis.[87,170]

E. Bilirubinemia

In some cases of benzene-mediated aplastic anemia bilirubinemia has been found. A mild bilirubinemia with the range of 1.3 to 1.8 mg/100 mℓ in 8 out of 28 patients belonging to our series was found.[87] On the other hand, we determined in 17 patients fecal urobilinogen by a semiquantitative method and found that nine had increased fecal urobilinogen ranging between 400 and 920 U/100 g of feces (normal: 50 to 300 U/100 g of feces).[87] It has been recently demonstrated that aplastic anemia and related disorders cause ineffective erythropoiesis in the bone marrow.[129,170] When this occurs, the denucleated red blood cells produced by the bone marrow are eliminated or the defective hemoglobin is broken down. The breakdown products of hemoglobin are ultimately converted to bilirubin and urobilinogen. Numerous experiments, especially those involving radioactive tracers, have shown this to be the case.[129,170] In order to differentiate between increased hemolysis or ineffective erythropoiesis as the cause of mild bilirubinemia, increased fecal and urinary urobilinogen, and reticulocytosis in some cases of benzene-induced aplastic anemia, erythrokinetic studies are required. If these changes are due to ineffective erythropoiesis, ^{15}N labelled glycine will appear in ^{15}N-labeled stercobilin in 10 days. The mild or moderate erythroblastemia found in some cases of benzene-induced aplastic anemia can be considered to have both a hemolytic component and a component that causes ineffective erythropoiesis.

F. Changes in the Osmotic Fragility

As it is known, the results of osmotic fragility tests in aplastic anemia are in normal range. Similarly, Goldwater[102] reported that a change in osmotic fragility is not frequently encountered in cases of benzene-mediated aplastic anemia. On the other hand, Erf and Rhodas[165] found an increase in osmotic fragility in only one of their cases with chronic benzene toxicity. A quantitative osmotic fragility test of Dacie[171] was performed in 20 out of 44 cases of aplastic anemia due to benzene toxicity before and after 24 hr of incubation at 37°C.[37,87,172] It was normal in seven of them. A mild or moderate increase either before or after incubation was detected in the remaining 13 patients. Kinkead et al.[173] performed an osmotic fragility study in dogs, rats, and monkeys exposed to 25 ppm of benzene or JP-4 vapors. JP-4 is a complex mixture of aliphatic and hydrocarbon compounds containing

0.3% benzene by weight. An increase in the red blood cell osmotic fragility in female dogs exposed to 5 mg/ℓ JP-4 was noted between 12 and 22 weeks of exposure.[173]

G. Benzene Toxicity and Erythrocyte Survival Time

Unfortunately, there are very few data in the literature on this subject. Paolino et al.[174] established a shortened life span of erythrocytes in animal experiments. The life span of the red blood cells was measured by the 51 Cr method in one patient with benzene-mediated aplastic anemia in our series and it was moderately shortened.[87] According to Goldstein,[39] "although usually anemia is primarily due to the failure of red cell production, the red cell life span may perhaps be somewhat shortened in many if not in most cases of benzene hematotoxicity". Despite the presence of several findings of increased hemolysis, the anemia is most commonly thought to be the result of decreased red cell production due to bone marrow depression.

H. Serum Lactate Dehydrogenase (LDH) Activity

As it is known, LDH in aplastic anemia is in normal range. Contrarily, this enzyme can increase in some patients with aplastic anemia due to chronic benzene toxicity. The present author and associates found in 6 out of 11 pancytopenic patients with chronic exposure to benzene that the serum lactate dehydrogenase activity is significantly elevated, ranging between 600 and 930 Wroblewski units.[87] In one patient, the LDH level returned to normal values following the improvement of the hematological state. In four patients with elevated LDH activity, there were findings of increased hemolysis, so increase in LDH activity may be attributed to the presence of a hemolytic component. On the other hand, in a pancytopenic patient in whom preleukemia later developed, the increased level of the activity of this enzyme may be explained by the latter state.[87]

I. Leukopenia, Secondary Infection, and Sepsis

The most frequently encountered finding of chronic benzene toxicity is leukopenia irrespective of the status of bone marrow, and it will lead to secondary infections and septicemia, particularly in advanced cases.[37] These complications were one of the main reasons for the fatalities in our series consisting of 44 patients with aplastic anemia due to chronic exposure to benzene;[37] 14 of them died either because of infections due to leukopenia or of associated hemorrhage due to thrombocytopenia.[37] Unfortunately, we obtained negative results in the culture of most of the cases. This failure of culture makes the therapy more difficult. In addition to neutropenia, abnormalities in leukocyte function have been emphasized by several investigators. First, Hektoen[175] reported that benzene toxicity is accompanied by a decrease in the production of specific precipitin and lysin. According to Hektoen, the generally increased sensitivity to infections that is observed in the cases of chronic benzene toxicity is due to three factors: (1) leukopenia; (2) deficiencies in antibody production; and (3) a decrease in their phagocytic properties. According to Koslova and Volkova,[120] a decrease in the phagocytic function may precede pancytopenia in mild benzene toxicity. On the other hand, Lange and associates[149] noted a decrease in the level of IgA and IgG in the individuals who were exposed chronically to benzene, toluene, and xylene whereas the level of IgM was normal. Additionally, the same investigators noted a decrease in the level of serum complement levels in 62 out of 79 inidividuals who were exposed to benzene or its homologues.[150] According to Smolik et al.,[150] the decrease in the immunoglobulin and complement levels are important from the viewpoint of pathogenesis.

J. Fetal Hemoglobin and Benzene-Mediated Aplastic Anemia

It is well known that the alkali-resistant hemoglobin, namely fetal hemoglobin, will begin to decrease after birth and will normally reach the normal level, which is 2% as assayed by alkali denaturation, when the baby is approximately 6 months.[129] An increase of hemoglobin F is usually found in beta-thalassemia, hemoglobinopathies such as sickle cell anemia, and in hereditary persistence of fetal hemoglobin.[129] An acquired increase of hemoglobin F may be observed in acute leukemia, in juvenile myeloid leukemia, and more frequently, in aplastic anemia.[176-179] The fetal hemoglobin increases to 3 to 15%, mostly in cases of aplastic anemia.[176] Levels above these are rare.

The present author and associates have determined fetal hemoglobin in 35 cases of aplastic anemia with chronic benzene toxicity.[37,87] The level of fetal hemoglobin ranged between 2 and 19.5% in 32 out of these 35 patients.[37,87,177,179] In three patients, the amount of fetal hemoglobin was very high, 30, 60, and 48%, respectively.[50,177] These three patients will be discussed separately. Bloom and Diamond[178] found a correlation between the prognosis in the cases of aplastic anemia and the amount of fetal hemoglobin measured. The prognosis is unfavorable in cases where the total fetal hemoglobin is less than 400 mg/100 mℓ, and better if it is higher. In our series of aplastic anemia due to chronic benzene exposure there were 16 patients with total fetal hemoglobin less than 400 mg/100 mℓ.[179] They all recovered. It is noteworthy that in 12 out of these patients the initial hematological state was severe.

Considering these facts, we suggested that in pancytopenic patients exposed to benzene, the total fetal hemoglobin value does not have the same prognostic value, as suggested by Bloom and Diamond,[178] in aplastic anemia. This point has also been emphasized by the present author and associates and Storti et al. in the patients with aplastic anemia due to various causes.[180-182] Shahidi and Diamond[183] and Aksoy and Seçer[176] have pointed out that the production of fetal hemoglobin excess is switched on in aplastic anemia and overproduction continues well beyond recovery. In several patients, elevated levels of fetal hemoglobin persisted for many years after recovery. Sometimes it rapidly returned to normal. This happened in one patient with aplastic anemia due to chronic benzene exposure in whom, after complete recovery, the fetal hemoglobin content rapidly dropped from 19.5 to 2%.[87] On the other hand, occasionally the mechanism is switched on a little later. In one pancytopenic patient with normocellular bone marrow, the fetal hemoglobin level was within normal limits when she was first seen. Following 3 months recovery the fetal hemoglobin level was found to be 8.7% whereas peripheral counts were within normal limits.[87] In three patients with aplastic anemia due to chronic exposure to benzene, the levels of fetal hemoglobin were as high as 33, 60, and 48%, respectively.[50,37,177] Two of them were brothers and heterozygous for high A$_2$ beta-thalassemia as well, this possibly being the reason for such a high finding.[50] One of these patients recovered and the fetal Hb content decreased from 60 to 8.5%. The second brother died from the complications of aplastic anemia. The third case of aplastic anemia with chronic benzene toxicity showed a high level of fetal hemoglobin, such as 48%, with an extremely low level of hemoglobin A$_2$ (0.7%).[177,184] Throughout his illness, the patient showed an unexplained hypochromic microcytic anemia.[37,177,184] Family studies which included one brother, one sister, and two children have given no evidence for the presence of a beta-thalassemia gene in his family or a gene for hereditary persistence of fetal hemoglobin. In this patient, the low level of Hb A$_2$ and marked elevated value for fetal hemoglobin were already evident during the pancytopenic period, 9 months preceding preleukemia.[37,177,184] This observation prompted us to suggest that low levels of hemoglobin A$_2$ with or without high levels of hemoglobin F, during a pancytopenic stage, may be an early sign of leukemia or preleukemia.[37,177,184] On the other hand, chemical characterization of hemoglobin F, assessing the numbers of glycine and alanine residue in γCB-3 was performed in five patients with benzene-mediated aplastic anemia.[177] In two patients, the glycine values in γCB-3 ranged between 0.36 to 0.83 residues.[177] In two patients, these values fell in the range of normal adults whereas the values of the other three were comparable to those found in newborns.[177] Recovery from aplastic anemia did not change these results.[177]

According to these results, glycine values in γCB-3 in aplastic anemia with chronic benzene toxicity vary from patient to patient.

K. Hemoglobin A₂ Level and Benzene-Mediated Aplastic Anemia

As it is known, the minor blood component, hemoglobin A_2, is composed of two alpha and two delta polypeptide chains. It usually constitutes 2 to 3% of total hemoglobin, but increases in beta-thalassemia syndromes.[129] It is decreased in δ beta-thalassemia, alpha-thalassemia syndromes, hereditary persistence of fetal hemoglobin, and Lepore hemoglobinopathies, etc. Aksoy and Erdem[185] were the first to note that it decreased in leukemia, namely, acute erythroleukemia. One year later, Weatherall et al.[186] reported the decrease of hemoglobin A_2 in another leukemic syndrome, the juvenile myeloid leukemia.

In 22 out of 28 patients with aplastic anemia with chronic benzene toxicity, hemoglobin A_2 values were found within normal limits.[37,87,177] Hemoglobin A_2 was definitely decreased in two patients, 0.67 and 1.5%, respectively, one of them mentioned earlier, and slightly in the other two, 1.9 and 1.95%.[37,87,177] In addition, in one patient hemoglobin A_2, like hemoglobin F, was within normal range on admission, but 3 months later, following a considerable hematological improvement, it decreased to 1.95%.[37]

L. Symptoms of Hemorrhagic Diathesis

The hemorrhagic manifestations of pancytopenia or aplastic anemia of chronic benzene toxicity are not different from those of other forms of aplastic anemia. Any part of the body can bleed. The most frequent are purpuric spots and echymoses. Digestive and urogenital system-associated bleedings or hematemesis are not uncommon. Hemorrhages in the digestive or cerebral system can be fatal. One of our patients had transient comas, most probably resulting from cerebral microhemorrhages.[87] The patient would become comatose for 24- to 48-hr periods, regaining consciousness completely again. The EEG revealed no abnormalities. In two of our cases, hemothorax or hemorrhagic pleuresy, a rarely encountered complication in aplastic anemia, developed.[187] Both patients died from this complication. A severe or moderate thrombocytopenia was found in all except four patients in our series consisting of 44 individuals.[87] One of the patients with a normal platelet level had severe anemia and leukopenia which were refractory to all treatment and his platelet levels, determined several times during a 1-year period, were always found to be within normal limits. In the blood smears of the patients with low levels of platelets, giant or morphologically abnormal platelets were detected.

M. Lymphopenia

Relative lymphocytosis is usually characteristic of aplastic anemia. Contrarily, in benzene-mediated aplastic anemia the absolute lymphopenia is not rare. It was observed in 24 out of 32 patients with aplastic anemia due to chronic exposure to benzene.[87] In the remaining eight patients, a relative lymphocytosis; lymphocytes ranging between 50 and 78% were found.

N. Monocytosis

In 4 of 32 patients with aplastic anemia due to chronic benzene toxicity, we detected a mild increase in monocytes, ranging from 9 to 13%.[87]

O. Pelger-Huet Anomaly and Benzene-Mediated Aplastic Anemia

In one patient with benzene-mediated aplastic anemia in our series, Pelger Huet anomaly had been detected following complete remission.[87] An incomplete family study did not confirm this inherited leukocyte anomaly, but the persistence of this finding for 2.5 years following his complete recovery is strongly in favor of true Pelger-Huet anomaly. Interestingly, this pancytopenic patient with a Pelger-Huet anomaly, showed a rapid clinical and hematological recovery. This observation prompted us to accept the possibility that such a congenital anomaly does not aggrevate the hematological effects of chronic benzene toxicity.[87]

P. Liver Tests

Selling[27,28] had reported paranchymal changes in the liver. These changes ranged from hepatitis to necrosis. Similarly, Wirtschafter and Cronyn[188] performed a study with a single benzene injection of 0.004 M/kg of body weight in rats. Following this procedure, they observed an increase in SGOT values and some cytoplasmic degeneration and mild nuclear changes histologically. Later, other studies like that of Rozera and associates[189] indicated that no histological changes occurred in the liver, although functional changes did take place. Rozera et al.[189] reported in animal experiments that there were abnormalities in the tests such as bromsulphalein, blood amino acid levels, and Takata-Ara functional assays. They also claimed that the liver disorders depend on the extent of hematologic changes as well as previous insults to the liver detoxification mechanism. In our series of 32 patients with benzene-mediated aplastic anemia, serum electrophoresis and liver flocculation tests, such as thymol turbidity and alkaline phosphatase levels, were determined. The results were within normal limits.[87] Only in one patient, the bromsulphalein test showed a 23% retention. In this patient the results of pyruvic and glutamic oxalacetic transaminases, thymol turbidity, prothrombin times, and those of proteinogram were within normal range. Following complete recovery, the results of the bromsulphalein test remained unchanged.

As it is known, occasionally viral hepatitis may cause aplastic anemia.[47,129] Hodgkinson reported five patients with chloramphenicol-hepatitis-aplastic anemia syndrome.[187a] Typically, the jaundice appears during or several weeks after a course of chloramphenicol, followed by the development of aplastic anemia.[187a] Recently we have observed a severe case of aplastic anemia associated with chronic benzene toxicity following 10 months after the appearance of hepatitis B.[187b] The propositus was a 32-year-old technician working for 6 years in a petroleum plant in eastern Thrace, Greece. A sample of the products obtained from this plant disclosed 2.2% of benzene, as determined by gas chromatography in the chemical engineering faculty of Istanbul University. As well, 2 years ago he had infectious hepatitis for a duration of 1 month. At that time he was not anemic and there were no signs of hemorrhagic diathesis. He never used chloramphenicol. Approximately 10 months later, epistaxis, lassitude, and purpuric spots appeared. He was admitted to the Haematology Department of Istanbul Medical School. He had severe pancytopenia (Hb 4.9 g/dℓ, PCV 0.17, platelets 54 $\times$ 10^9/ℓ, WBC 2.2 $\times$ 10^9/ℓ) with 6% neutrophilic band forms, 27% segmented neutrophils, 6% monocytes, 1% eosinophils, 1% basophils, and 59% lymphocytes. The bone marrow aspirate was very hypocellular, and the findings of the bone marrow were consistent with the diagnosis of severe aplastic anemia. Proteinogram disclosed hypogammaglobulinemia (γ-globulin 8.6%). The liver tests were in normal range. HBs Ag and HBe Ag were negative and anti-HBs, anti-HBe, and anti HBc were positive. These tests were performed by the enzyme-linked immunosorbent assay (ELISA) method in the Department of Microbiology, Istanbul Medical School. The patient is still under therapy. In the case of this technician who was exposed to benzene for 6 years, 10 months after disappearance of hepatitis B, a severe aplastic anemia developed. We are strongly in favor of proposing that viral hepatitis acted as a contributary factor in the development of severe aplastic anemia associated with chronic benzene exposure. Therefore, the individuals who might have been exposed to benzene, should be protected from viral hepatitis by measures such as vaccination.

Q. Nervous System and Chronic Benzene Toxicity

The effects of benzene on the central nervous system were described earlier in the chapter. There are few reports concerning the effect of chronic benzene toxicity on the nervous system. Truhaut[190] had indicated the possibility of long-term effects of benzene on the nervous system such as polyneuritis of lower extremities. In 1973, Kahn and Muzyka[75] observed that in 74 workers with benzene exposure, there were complaints such as frequent headaches that appeared towards the end of the shift, lassitude, easy exhaustion, and disturbances in sleep and memory. Also, 60% of the workers examined had increased δ-aminolevulinic acid in erythrocytes and 30% of them also had increased coproporphyrine in urine. On the other hand, there was no exact relationship between the duration of exposure to benzene and the extent of the disorders in porphyrine metabolism. Additionally, it was established that all workers with increased ALA values in erythrocytes had functional disorders of the nervous system. According to the investigators, the abnormalities of porphyrine metabolism caused by chronic benzene toxicity took place before the appearance of morphological changes in peripheral blood.[75] Determination of ALA in the erythrocytes and abnormalities in porphyrine metabolism are the most sensitive indicators of the toxic effect of benzene. Additionally, a good correlation exists between the disorders of porphyrine metabolism and functional disarrangements of the central nervous system. Finally, early individual sensitivity of porphyrine metabolism in workers exposed to benzene may have a prognostic importance and may be used for prophylactic purposes.[75]

On the other hand, Kahn and Muzyka[75] experimentally showed that chronic benzene toxicity has some effects on the porphyrine metabolism in rabbits. In this study, they have found that benzene toxicity causes changes in the levels of δ-aminolevulinic acid, porphobilinogen, and coproporphyrine in the central nervous system.[75] In spite of clinical and experimental findings, they did not study the peripheral nervous system.

Considering the above mentioned facts, Baslo and present author[53] have performed neurological, electromyographical, and motor conduction velocity examinations in six patients with benzene-mediated aplastic anemia, and two cases of preleukemia with chronic benzene toxicity. In addition, sensory conduction velocities were measured in three patients. Neurological abnormalities were observed in four out of six patients with benzene-mediated aplastic anemia. One patient had global atrophy of the right leg. In three patients, deep reflexes of both sides were exaggerated and in one, distal, superficial sensory disturbances were detected on upper extremities. The results of electromyographic (EMG) and conduction velocities (motor and sensory) showed that in one case there were global atrophy and decreased sensory vibration of lower extremities.[53] This patient had global atrophy of the right leg. While his aplastic anemia improved, the EMG neurogenic involvement with denervation showed regression, at the same time motor conduction velocity (MCV) values increased. This case was exposed to the toxic substance for 7 years. Sensory conduction velocity of the peroneal nerve was normal when he recovered. In this patient, global atrophy and decreased MCV suggested the presence of neurogenic involvement due to anterior horn and/or peripheral nerve lesions.

In another case, EMG revealed a neurogenic involvement and MCV of peroneal and tibial nerves were normal, but MCV of the median nerve showed distal latency lengthening. In this case slight distal neurogenic involvement of upper limbs was detected. The patient had been exposed to benzene for 6 years before performance of neurological examination. Another case showed normal EMG and MCV of the peroneal nerve, but MCV of median and ulnar nerves showed distal latency lengthening. These findings suggested slight distal neuropathy of upper extremities. He was exposed to benzene for 8 years with the exclusion of the last 53 months. Neurogenic findings were not severe since the patient was not exposed to this toxic agent for a long period.

Another patient with normal neurological results had been exposed to benzene for 6 years and the exposure ceased 96 months earlier. These normal results were probably due to the long period of nonexposure. In another patient, despite 8 years of exposure, except the last 4 months, the results of neurological examination were normal. In this case it is difficult to explain these normal results. They could be attributed to less exposure to benzene rather than to the short period of nonexposure. Moreover, this patient recovered hematologically in a short time. In another patient, the results of EMG examination and motor conduction studies were normal, but a decrease in the sensory conduction velocities of the lower extremities was observed. In particular, the amplitudes of nerve action potentials were low. Briefly, four out of six cases of benzene-mediated aplastic anemia showed neurologic abnormalities, but two were normal. These findings prompted the investigators to suggest that benzene may show toxic effects on the nervous system at the level of peripheral nerves and/or spinal cord.

Polyneurotis to n-hexane is well established by the studies of Ikeda and Herskowitz.[191,192] The absence of this industrial compound in the adhesives and benzene solutions used by workers involved, eliminates this possibility of n-hexane polyneuritis. On the other hand, similar changes in the nervous system were observed by an experimental study by Matsumato[193] on chronic toluene toxicity in rats. In this study, electric threshold velocities (motor nerves) were measured after chronic exposure to toluene. An increase was noted in the ratio of threshold stimuli, and a decrease in the maximal conduction velocity. These changes were attributed to the influence of toluene on the nervous system. Contrary to n-hexane, the role of toluene in the development of neurological abnormalities described in the patients of Baslo and Aksoy[53] cannot be eliminated easily because of the presence of toluene, although at a low level ranging between 6.37 and 9.25%. Here we would like to emphasize the generally accepted fact that in aplastic anemia, changes in the nervous system are found only when hemorrhage occurs there.[129]

R. Splenomegaly

The absence of splenomegaly is essential for the diagnosis of aplastic anemia.[129] When it is present, another disease other than aplastic anemia such as leukemia or myeloid metaplasia, etc. should be seriously considered. In contrast to this, in several autopsy reports dealing with aplastic anemia due to chronic benzene toxicity, the development of the splenomegaly have been recorded.[194] Zayer[194] reports that splenomegaly is frequently encountered in the autopsies of cases involving benzene, although it is rarely found as a clinical manifestation. In some of these cases the presence of hematopoietic activity or fibrosis in the spleen are noted. Mostly these cases present the findings of myeloid metaplasia or those of leukemia due to chronic benzene toxicity. Splenomegaly was observed during the course of the illness in 2 out of 44 pancytopenic patients with chronic benzene toxicity.[37,87] In one patient, a marked hepatosplenomegaly, causing very severe abdominal pains, had developed nearly 2 weeks before death and possibly the cause of splenomegaly was the development of acute leukemia. In the second case, the patient showed the findings of aplastic anemia with hypoplastic bone marrow. But in this patient, during the course of his illness, giant erythroid precursors comprising 38 to 72% of the nucleated cells in a hyperplastic bone marrow (in three bone marrow punctures) were found. In this case, the cause of hepatosplenomegaly was preleukemia which developed in the course of aplastic anemia with chronic benzene toxicity.[87]

S. Bone Marrow

Chronic benzene toxicity can result in extremely variable changes in the bone marrow. The picture can range from complete aplasia to a highly hyperplastic bone marrow. In some of the patients, the bone marrow can be found fully acellular in the terminal stage. There are practically no hematopoietic cells left. Three of our cases of aplastic anemia with chronic benzene toxicity were fully aplastic and acellular.[87] Even repeated marrow punctures failed to yield a bone marrow material in one case. Contrary to this, Mallory et al.[101] reported that most of their benzene-mediated aplastic anemia cases exhibited hyperplastic bone marrows.

Also, 6 out of 19 cases studied showed hypoplasia, two were normocytic, two were leukemic, and the remaining nine were hyperplastic. The hypoplastic bone marrows were deficient in cells, with increased fat. They also had cell clumps; $^2/_3$ of them were erythroblasts. The cells of the granulocytic series were usually less than the erythroblasts. Even through hyperplasia was evident, stem cells that resembled "megaloblasts" were observed. In one case, mono- or polynucleated plasma cells had increased, although erythroblasts were the most frequently encountered marrow element. A decrease in the erythroblastic series and a relative increase in the granulocytic series will occur in a certain number of cases.[101] The majority are myelocytes and they are more numerous than the mature neutrophils. According to Mallory et al.,[101] fibrous tissue, rather than fatty tissue, replaces hematopoietic tissue.

Mallory et al. suggested that the effects of benzene on the bone marrow will vary from individual to individual and are independent of length of exposure. More than one type of foci may be encountered in the bone marrow. For example, a hypoplastic focus may be seen next to a hyperplastic focus. Penati and Vigliani[195] have classified the bone marrow picture in chronic benzene toxicity into:

1. Complete or partial aplasia in the bone marrow.
2. Pseudo-aplastic anemia, with the marrow being hyperplastic.
3. Atypical anemias, leukocytosis, and splenomegaly.
4. Benzene-mediated leukemia.

On the other hand, according to Rohr,[196] the first symptoms of chronic benzene toxicity can be an increase in the myelocytes, with a decrease in the granulation and an increase in the vacuolization in the bone marrow. In one of these cases, Rohr found that the granulocytic series cells were decreased, but reticulum cells and macrophages had become more abundant. In some cases plasma cells will increase. The most important change in the bone marrow that accompanies hyperplasia and hypoplasia, is the decrease in maturation. Although myelocytes and metamyelocytes increase, neutrophils, that is the mature granulocytic series cells, are very rare. The same situation is frequently encountered in the erythrocytic cell maturation pathway. Although mature orthochromatic normoblasts are infrequent, polychromatic normoblasts and basophilic erythroblasts can constitute the majority. Steinberg[197] found a decrease in cell maturation and division at the immature reticulum cell level in animal experiments exposed to benzene. Rondanelli et al.[198] performed a cinomicrophotographic study in three patients with benzene erythropathy and in one with erythroleukemia due to occupational exposure to benzene showing changes which can be summarized as follows. Morphological anomalies of erythroblasts in interphase: the interkinetic nuclei of erythroblasts may be atypical in form (irregular), volume (gigantism, or microerythroblasts), and structure (bi- and plurinucleated cells and structural abnormalities). Chromosomal changes found in newt experiments will be summarized in Chapter 6.

On the other hand, Moeschlin and Speck[199] studied the bone marrows of animals that were poisoned with benzene by autoradiography with ^3-H methyl thymidine, and found that the results varied from animal to animal. They examined 19 bone marrows and found that 4 were very hypoplastic, 4 were hyperplastic, 5 were normocellular, and 4 were hypercellular. No correlation was found between cellularity of the bone marrow and the duration of exposure to benzene. The decrease in the myeloid series precursors was particularly striking. Although the erythropoietic series had fared relatively better, some qualitative differences were present. Numerous changes in chromatin structure had occurred. A left shift in erythropoiesis was present. Polynucleated, underdeveloped cells (erythrogons) were other characteristic findings. The abundance of erythropoietic series cells is a striking feature of benzene-mediated aplastic anemia cases that exhibit a hypercellular bone marrow.

Sometimes cells of the granulocytic series may increase also, which can be considered a preleukemic symptom, as we shall discuss in Chapter 6. Moeschlin and Speck[199] called this benzene-mediated hypercellularity, pseudohypercellularity. The hypercellularity, particularly in erythroblastic series, in some benzene-mediated aplastic anemia cases are in contrast to the definition of aplastic anemia, hypocellularity of bone marrow, without infiltration by neoplastic cells or significant fibrosis, being essential.[200] In contrast, according to Wintrobe et al.,[129] "In still others, but especially in those with benzene and in some who have had internal or external irradiation, the bone marrow has been found hyperplastic." Indeed, in a study performed by the present author and associates in which 44 pancytopenic patients with chronic benzene exposure were followed for 2 to 17 years, only 21 patients had hypocellular bone marrow.[37] Bone marrow was normocellular in 13 of the patients and hypercellular in 8 (bone marrow aspiration was not performed in two patients). Again, according to Wintrobe et al.,[129] "It is recognized that the morphologic structure of the bone marrow, particularly as seen in a single small aspiration or biopsy sample, does not necessarily reflect an accurate total bone marrow function and that a clinical and blood picture resembling the acute disorder described by Ehrlich or even the more chronic forms, will not always be accompanied by a completely fatty marrow". Sampling errors or functional "hot pockets" may also contribute to the lack of correlation between functional capacity and cellularity of bone marrow.[201] Therefore, hypoplastic bone marrow is not a sine qua non for the diagnosis of aplastic anemia.

As we emphasized above, this situation is particularly true for aplastic anemia due to chronic exposure to benzene. On the other hand, sometimes the proportion of the erythroblastic series elements can increase in a fashion that is reminiscent of hemolytic anemias. In one case described by Custer,[202] the cells of the erythroid series constituted 90% of the elements of the bone marrow. In this case the possibility of an acute erythroleukemia due to chronic benzene toxicity should strongly be considered. Sometimes elements of myeloid series increase also which can be considered to be a preleukemic state. This possibility will be discussed in Chapter 6. As we mentioned above, according to Moeschlin and Speck[199] this benzene-mediated hypercellularity is in fact a pseudohypercellularity.

With regard to our results in our series of aplastic anemia with chronic benzene toxicity, some features of them are mentioned in this chapter. A maturation arrest in the granulocytic and erythroblastic series and erythroid hyperplasia were the most commonly encountered finding in our pancytopenic patients with chronic exposure to this chemical.[37,87] The elements of erythropoietic series were increased, ranging between 36 and 77%, in a total of 42 bone marrow.[37,87] Twenty of them had maturation arrest in myeloid series whereas fifteen of them showed the findings of maturation arrest in erythropoietic series.[37,87] In five patients, a mild or moderate megaloblastic erythropoiesis was found.[37,87] In one of them, megaloblasts ranged between 1 and 22%, accompanying normoblastic erythropoiesis.[37,87] The very marked erythroblastosis found in the peripheral blood smear of two other cases was eliminated following folic acid therapy. Mallory et al.[101] had previously shown that a megaloblastic character may be associated with some of the erythroblasts in benzene-mediated aplastic anemia.

Megaloblastic erythropoiesis in refractory anemia with a hyperplastic bone marrow is considered to be due to a block in nucleoprotein formation.[203] Furthermore, folic acid deficiency may also play a role in the development of megaloblastic erythropoiesis observed in some pancytopenic patients exposed to benzene. In one of these five patients, the low level of serum folic acid confirmed this possibility.[87] However, an unfavorable response to folic acid therapy showed that folic acid deficiency was not the only factor in the development of a megaloblastic erythropoiesis in that patient. Furthermore, in another patient with megaloblastic erythropoiesis, the disappearance of a marked erythroblastemia after folic acid therapy may be considered as confirmation of folic acid deficiency, although the folic acid level was not determined.[87] Thus, in the development of megaloblastic erythropoiesis in the patients exposed to benzene, at least two factors may play a role: (1) a block in the nucleoprotein formation; and (2) folic acid deficiency. On the other hand, in two patients in our series, giant erythroid precursors which were found to vary between 9 and 72% of nucleated cells were found.[87] In one of them, erythroid precursors appeared during the course of the disease. In this patient, four bone marrow punctures were performed.[87] The first bone marrow findings were consistent with the diagnosis of aplastic anemia due to chronic benzene toxicity. The bone marrow was hypocellular and there was a maturation arrest in erythroid series. Erythroid cells resembling giant erythroid precursors were not found. Some months later, when the general status of the patient deteriorated, three other bone marrow punctures were performed. In three myelograms, the bone marrow was hypercellular and the so-called giant erythroid precursors appeared ranging between 38 and 72%.[87] They were generally encountered in clumps. They seemed to be very immature cells with large amounts of cytoplasm that was heavily basophilic. The chromatin in the nucleus was generally thin, with three to four heavily stained nucleoli visible. In some of them, the cytoplasm was basophilic as plasma cells. Nuclei were sometimes doubled and frequently atypical. Some of these cells looked like promegaloblasts. They often had one to four large nucleoli and showed numerous mitotic forms. Unstained areas were also found in the cytoplasm of these giant erythroid precursors. These cells were similar to those described by Mallory et al.,[101] and erythrogens observed by Moeschlin and Speck[199] in animals experimentally poisoned by the injection of benzene. In the majority of the patients, a vacuolization of the myeloid, erythroid, and plasma cells were observed in our series.[87]

T. Electron Microscopic Findings in Benzene-Mediated Aplastic Anemia and Leukopenia with Chronic Benzene Toxicity

Erbengi and the present author[204] performed ultrastructural studies in two cases of benzene-mediated aplastic anemia and in two leukopenic individuals with chronic benzene exposure. Three of them were shoe workers and one handbag maker with a duration of exposure to benzene varying between 4 and 17 years (mean: 8). Ultrastructural findings were in accordance with changes in the bone marrow observed by light microscopy. In these two patients with leukopenia studied with light microscopy, bone marrow was normocellular, associated with maturation arrest in the myeloid elements. Ultrastructural observations of both cases were in accordance in the above mentioned findings. Additionally, there was an increase in plasma cells with maturation arrest in the later phases of erythroid series.

Electron microscopic studies shed some light on the phagocytic activity of the reticulum cells. This phenomenon was observed in two patients, one with aplastic anemia and the second with leukopenia. On the other hand, ultrastructural studies showed a hyperactivation of erythropoiesis in the bone marrow in one case of aplastic anemia and a depression in the erythroid elements in the second case. These different findings may give the impression that the effect of benzene on the bone marrow may change from patient to patient.

V. Pathology

As expected, the main pathological changes associated with cases involving benzene toxicity relate to hematopoietic tissue. Therefore, we will emphasize the pathological findings of the bone marrow. Bone marrow: findings in the autopsies of the victims of chronic benzene toxicity are quite variable. All types of findings ranging from a totally acellular bone marrow to hypercellular or even a leukemic marrow are possible. The bone marrow frequently manifests different types of changes at the same time, in different foci. There might be a totally hypocellular focus right next to a hypercellular one. Some researchers believe that complete aplasia of the bone marrow is an extraordinary occurrence. For example, Mallory and associates[101] were unable to find aplasia of the bone marrow in any of the cases they studied. Fat content increases radically in hypoplastic marrow. Mallory et al.[101] reported that this causes the blood-producing regions of the bone marrow to resemble yellow marrow.

The sinusoids are rarely filled with blood, frequently being empty and collapsed. Intrasinusoidal foci, mainly consisting of erythropoietic elements, may be seen among them. Latta and Davies[109] demonstrated that the myeloid tissue of experimental animals is stimulated at the onset, and the bone marrow is hyperplastic. As the study is continued, changes that would indicate degenerative changes in the myeloid tissue will be observed and will slowly disappear. Erythropoietic tissue is more resistant to degeneration, and changes occur later. Maturation is arrested and stem cells will degenerate; they become transformed to macrophages and eventually, hypoplasia in erythropoietic tissue occurs.[109] The dense basophilia of the hemoblast cytoplasm and the clumped, pale, or achromatic nature of the chromatin in the nucleus are striking findings.[109] The existence of regenerating foci, next to degenerative changes in the bone marrow, are known from animal experiments since the time of Selling.[19,27,28] The variable results obtained from animal experiments are particularly due to differences in the experimental conditions.

X. Spleen

As we emphasized above, although splenomegaly is a rare occurrence in chronic benzene toxicity, it is frequently found in autopsies.[194] Mostly these are either due to presence of leukemia or that of myeloid metaplasia. The findings of these two pathologic conditions will not be discussed here. Free macrophages with high hemosiderin that mainly phagocytize erythrocytes are abundant in the spleen. Many researchers have drawn attention to the hyperplasia encountered in the reticuloendothelial cells. Decreases in lymphocytic tissue, and hyperplasia in fibrotic tissue have been reported in animal experiments.[109] Increases in reticulum cells and macrophages have also been reported.[109]

Y. Liver

Changes in the liver resemble those in the spleen and the bone marrow. Kuppfer cells are always evident. They are generally filled with hemosiderin. Central necrosis is frequently encountered. This last observation has not been totally corroborated yet. Erythropoietic foci may be detected in the cases that involve myeloid metaplasia.

Z. Lymphnodes

The lymphnodes do not seem enlarged in clinical studies. In autopsies they give the impression of having been enlarged. The lymphocytes in the lymph cords are usually decreased. According to Mallory et al.,[101] erythroblast and monocytes can be found in lymphnodes, possibly due to myelosclerosis. Animal experiments indicate a decrease in lymphnodes and lymphoid tissue.[109] Changes of secondary nature can occur in other organs outside hemopoietic tissue. Some of these changes occur in the endocrine glands, and none of them are very important. A depression and retardation in the maturation of the germinal epithelium of the testes have been reported.

AA. Prognosis and Mortality

The prognosis in benzene-mediated aplastic anemia depends mostly on the extent of the changes in the bone marrow. If the pancytopenia is diagnosed early, that is to say, recognizable before anemia, leukopenia and particularly, thrombocytopenia become severe, and if a suitable therapy is initiated, the prognosis will be favorable. Contrarily, the patient's chances mostly decrease once pancytopenia is established and secondary infections and hemorrhagic diathesis occur. The prognosis will be deteriorated when the patient continues to be in contact with benzene after diagnosis and warnings. Although there are numerous reports on the occurrence of pancytopenia associated with chronic exposure to benzene, only few papers dealing with the problem of outcome among pancytopenic patients with chronic benzene toxicity appeared in the literature. Hunter[140] reported 10 fatal cases of a total 89

Table 4
**RELATIONSHIP AMONG AGE, DURATION OF EXPOSURE, AND
OUTCOME INCLUDING DEVELOPMENT OF LEUKEMIA IN 44
PANCYTOPENIC PATIENTS ASSOCIATED WITH CHRONIC
BENZENE TOXICITY**

		No. of patients	Age (years)	Duration of exposure (months)
Pancytopenia	Complete remission	23 (52.3%)	26.87 ± 9.50	65.43 ± 42.30
	Fatal outcome	14 (31.8%)	37.0 ± 9.81	93.0 ± 41.49
	Leukemia developed	6 (13.6%)	35.0 ± 7.46	94.67 ± 60.17
	Myeloid me-taplasia developed	1 (2.3%)	35	96
	Overall fatal outcome	21 (47.7%)	36.0 ± 8.98	93.62 ± 45.0

From Aksoy, M. and Erdem, Ş., *Blood,* 52, 285, 1978. With permission.

pancytopenic patients with chronic exposure to this aromatic hydrocarbon. In Germany, 164 cases of benzene-induced aplastic anemia were encountered between 1913 and 1928;[205] 27 of them were fatal.[205] This corresponds to a mortality rate of 16.4%. Six of seven cases studied by Nilsby were fatal.[206] Contrary to this, in another report from Sweden, only 2 of 60 cases were fatal.[169] On the other hand, Vigliani and Forni[207,208] studied 83 patients with benzene-induced hematoxicity in the provinces of Milano and Pavia; all of 50 patients who were still alive at that writing had aplastic anemia, and 19 patients with acute leukemia and 14 with aplastic anemia in this series ended fatally. Mortality rate among the patients with aplastic anemia was 20.2%.

As can be seen from Table 4, in our series consisting of 44 pancytopenic patients with chronic exposure to benzene, the overall fatal outcome was 47.7%;[37] 31.8% of them died from the complications of pancytopenia. In 13.6% of them, leukemia developed later. One patient (2.3%) was lost from myeloid metaplasia. In this patient, myeloid metaplasia developed in 8 years after recovery from benzene-induced aplastic anemia.[37,134] Seven pancytopenic patients died (15.9%) in the first year following diagnosis, six in the second year (13.9%), and only one (2.3%) died after 5 years of survival.[37] In this series, as can be seen from Table 4, there was a relationship among some clinical findings such as age, duration of exposure, and outcome of pancytopenic patients with chronic exposure to benzene. The mean age of pancytopenic patients with complete remission was significantly less than those with overall fatal outcome, 26.87 and 36 years, respectively.[37] Similarly, the mean duration of exposure was significantly shorter in the pancytopenic patients who recovered completely than those with overall fatal outcome, 65.43 and 92.62 months, respectively. Despite this, in one of the patients in whom preleukemia developed 6 years after complete recovery, the duration of exposure was very short, only 4 months. But this patient was heavily exposed to benzene during this short period. However, there was no relationship between the data concerning pancytopenia such as hemoglobin, hematocrit, leukocytes, platelets, and the outcome, including leukemia, in our series.[37] Here the only exception was the mean values of neutrophils in peripheral blood. In cases of aplastic anemia with complete recovery, the mean values of neutrophils were significantly higher than those found in overall fatal outcome.[37] Contrary to the findings of peripheral blood, there was a clear relationship between

Table 5
THE RESULTS OF CYTOGENETIC STUDIES OBTAINED IN 10 PANCYTOPENIC PATIENTS WITH CHRONIC BENZENE TOXICITY

Case No. and (age)	Duration of exp. (years)	Numerical	Chromosomal aberrations
1 (26)	10	Polyploidy 2%	Normal
2 (29)	4	None	Secondary constrictions, chromatid inter-changes, achromatic lesions, dicentric chromosome
3 (16)	4	None	Breakage, chromatid interchanges, acentric fragment
4 (34)	7	None	Normal
5 (23)	10	47 16%	Normal and abnormal clones hyperdiploid karyotypes, extra (C), chromatid separation
6 (31)	5	None	Normal
7 (30)		Hypodiploid 45 11% Hyperdiploid 47 12% Polyploid 88 92%	Dicentric chromosomes, normal, abnormal clones, ring chromosomes, endoreduplication, extra (C)
8 (44)	7/12	None	Normal satellite association
9 (57)	3	47 16%	Breakage, secondary constriction, dicentric chromosome, acromatic lesion, extra (C)
10 (39)	10	None	Dicentric chromosome, acentric lesion

From Erdoğan, G. and Aksoy, M., *New Istanbul Contrib. Clin. Sci.*, 10, 230, 1973. With permission.

the types of the cellularity of bone marrow and outcome, including development of leuke-mia.[37] Of 21 pancytopenic patients with hypocellular bone marrow 11 died (52.4%) from the complications of aplastic anemia and in five patients (23.8%), leukemia developed later. Contrary to this, only 1 of 13 pancytopenic patients with normocellular bone marrow (7.7%) died. On the other hand, four out of eight pancytopenic patients with hypercellular bone marrow recovered completely (50%) and two died from complications of pancytopenia (25%). In one of these two, leukemia developed 1 year later and in the other, the occurrence of myeloid metaplasia 9 years after complete recovery was the cause of death.[37,134] There was no relationship between the levels of hemoglobin F and the outcome of the pancytopenic patients with chronic benzene toxicity.[37,87,177,179,184] Increased levels of Hb F, with or without decreased levels of Hb A_2, were found in pancytopenic patients, both with complete remission and fatal outcome, but very high levels of Hb F and very low levels of Hb A_2 are encountered only in leukemic states[37,87,177,184] unless the presence of a beta-thalassemic gene or a gene for hereditary persistence of fetal hemoglobin is eliminated.[37,87,179,184]

In one of our patients, a very low level of Hb A_2 and high level of Hb F had been found already during the preceding pancytopenic stage whereas leukemia developed 9 months later.[37,177,178,184] This observation was mentioned earlier.

On the other hand, in 10 out of 44 pancytopenic patients with chronic benzene toxicity, cytogenetic studies were performed by Erdoğan and the present author.[209] Six of them had complete remission, two showed fatal outcome, and in two leukemia developed later (Table 5). In these cases there was no relationship between cytogenetic findings and outcome, including the development of leukemia following benzene-induced pancytopenia.

Hernberg et al.[210,40] have performed a reexamination study in 125 workers who were occupationally exposed to benzene 9 years before this study. At that time more than half showed severe or moderate findings of chronic benzene toxicity.[40] Hematological analysis in these individuals showed that the prognosis of the severe case did not differ from that of the mild ones, provided the acute stage had been passed. One patient developed leukemia after a latency of 7 years whereas most of the other individuals chosen associated with severe findings of chronic benzene have recovered.[210]

As it is known, before bone marrow transplantation the mortality rate of aplastic anemia cases were generally high.[129] According to some investigators, it is between 55 and 74%.[211,212] In our series of 44 pancytopenic patients with chronic benzene toxicity, the complete remission was 52.3% and overall fatal outcome, including the development of leukemia and myeloid metaplasia, was 47.7%.[37] This result is slightly more favorable than the mortality rates obtained in the series of acquired aplastic anemia of various etiology. It is useless to emphasize that in these patients with benzene-mediated aplastic anemia modern therapeutic methods, such as bone marrow transplantation and antilymphocytic serums, were not used.

BB. Pathogenesis of Benzene-Induced Aplastic Anemia

Although the pathogenesis of aplastic anemia due to chemicals and drugs is not completely understood, there are four possible mechanisms which are mainly forwarded for the explanation of this hematologic disorder: (1) absence or defects in pluripotent stem cells (stem cell deficiency); (2) microenvironment deficiency; (3) abnormalities of hemopoietic regulatory cells or factors (a defective control mechanism); and (4) viruses.

As it is known, exposure to benzene has been implicated in the development of several types of blood dyscrasies such as leukopenia, thrombocytopenia, pancytopenia, leukemia, malignant lymphoma, etc. Involvement of different hemopoietic cell lines in benzene toxicity was the reason in suggesting that the main target of benzene or action of its metabolites may be pluripotential stem cells. Despite this, early experiments were not in favor of this assumption that benzene or its metabolite action may be pluripotential stem cells, and there are some studies which indicate that the multipotential stem cells are relatively resistant to the toxic action of benzene.

As early as 1949, Steinberg[213] found a decrease in cell maturation and division at the immature reticulum cell level in experiments in rabbits intoxicated with benzene. According to Steinberg, at least one part of the mechanism by which benzene induces aplasia of bone marrow is probably the inhibition of cell division and maturation past the level of primitive reticular cell. Furthermore, Moeschlin and Speck,[199] and later, Kissling and Speck[214] studied the mechanism by which benzene affects the bone marrow using autoradiographic techniques with ^{3}H-thymidine and ^{3}H-cystidine in rabbits. They found that benzene inhibits DNA synthesis, causing pancytopenia. An inhibition of the development occurs at the basophilic normoblasts in the erythroid and the promyelocyte in the myeloid series. The investigators interpreted their results as showing a marked myelotoxicity resulting from inhibition of DNA synthesis. But as Freedman[215] pointed out rightly, "however, these studies do not prove that the primary toxic event is indeed the inhibition of DNA synthesis, as these studies were performed after toxicity was manifest."

Furthermore, Kissling and Speck[214] have performed experiments in mice given 300 mg of benzene per kilogram of weight per day, subcutaneously for 2 weeks. At the end the study, radiographic investigations of the bone marrow in vivo were performed by ^{3}H thymidine. The results of this study indicated that the hematopoietic stem cells assayed by colony forming techniques were intact.[216] Speck and Kissling[216] concluded that the myelotoxicity of benzene is due to interference with DNA synthesis of bone marrow cells and not to damage to hematopoietic stem cells. But again, as Freedman pointed out, "supportive data for this concept have not been forthcoming."[215]

Hemopoietic stem cell has at least two important capabilities: (1) self-renewal, (2) differentiation to produce a variety of lineage-restricted progenitor cells. In other words, long-term maintenance of the hemopoietic system in vivo depends on the primitive stem cell to undergo self-renewal and differentiation. Any block or disturbance in these two capabilities will cause either aplastic anemia (a block in self-renewal) or leukemia (a block in differentiation).[216a]

On the other hand, Uyeki et al.[217] showed in 1978 that when mice were submitted to inhalation of benzene vapor (4680 ppm) for 8 hr, a significant depletion in bone marrow colony-forming cells (CFC) was observed in the cultures in vitro assayed 1 day after benzene inhalation. Thus, Uyeki et al.[217] concluded that precursors of the hemic cell renewal system are sensitive to benzene inhalation in mice.

Several schedules of benzene exposure were evaluated for their effects on peripheral white blood cell counts, bone marrow cellularity, and transplantable colony-forming units (CFU-S) in mice by Gill et al.[218] Intermittent exposure to 4000 ppm benzene in air produced leukopenia without altering bone marrow cellularity. Yet, the same treatment, however, decreased the number of CFU-S to 30% of control values. Uninterrupted exposure to lower levels of benzene (such as 100, 500, or 1000 ppm) decreased peripheral cell counts within 24 hr, and later decreased marrow cellularity. Exposure of a nondividing population of stem cells (CFU-S) to benzene for up to 24 hr produced no detectable effect on the subsequent development of spleen colonies, suggesting that the effect of benzene on CFU-S occurs only after peripheral cells are depleted. According to Gill et al.,[218] these findings indicate that benzene has effects on both differentiated cells and undifferentiated stem cells. Despite this, Gill et al. forwarded that an effect on the pluripotential stem cell is an important aspect of benzene toxicity, but not its exclusive or initial site of action.

The hematotoxicity and possible leukemogenicity of benzene was tested by inhalation in mice by Horigaya et al.[219] Groups of mice were exposed either to normal air or air containing benzene of 4000 ppm for 6 hr/day, 5 days/week for 9 days or 11 consecutive days. Cultures were made from bone marrow samples. Each culture contained an adherent fraction, consisting of phagocytic mononuclear cells, fat cells, endothelial cells, and epithelial cells and a nonadherent fraction, consisting mainly of granulocytes and macrophages. The number of spleen colony-forming cells was progressively reduced in cells from benzene-treated mice. Cultures from benzene-treated mice had a reduced capacity to maintain stem cell proliferation compared to cell cultures from normal cells, as shown by combining normal or treated adherent fraction with various reinoculations. According to Horigaya et al.,[219] benzene injures stem cells, reducing their ability to self-replicate and dearranging the adherent cell population. Neonatal mice injected with cells from benzene-treated mice did not develop leukemia during 8 months after injection. The investigators concluded that benzene may promote, rather initiate leukemia.

Furthermore, Tunek et al.[220] injected into mice benzene and its metabolites such as phenol, hydroquinone, and 1,2-dihydro-1-1,2-dihydroxybenzene. The adverse effects of benzene on the concentration of the bone marrow cellularity in tibia were measured. Benzene showed strong toxic effects. Six daily injections of 440 mg/kg of benzene reduced cellularity and colony-forming units per tibia by 86 to 95%. None of the benzene metabolites tested showed similar effects of benzene. Despite this, phenol affected slightly but significantly the stem cell concentration. Toluene, a competitive inhibitor of benzene metabolism significantly reduced the effects of benzene. Tunek et al.[220] reported that regeneration of bone marrow after benzene injections occurred rapidly during the first week, then at a slower rate for the next 4 weeks.

Baarson et al.[221] exposed mice to 10 ppm benzene for 6 hr/day, 5 days/week. This trial caused a progressive depression in vitro colony-forming ability of one the progenitor cells (CFU-E). Colony growth of cells from exposed mice was only 5% of control colony growth after 178 days of exposure. Burst-forming cell growth was depressed to 55% of control after 66 days, but returned to control growth values at 178 days. Additionally, benzene-exposed mice exhibited depressions in the number of splenic nucleated red cells and in the numbers of circulating red blood cells and lymphocytes. According to Baarson et al.,[221] these results indicate that low-level exposure to benzene may be hematotoxic.

On the other hand, there are very few data showing the role of microenvironment deficiency in the development of benzene-induced aplastic anemia. Frash et al.[222] showed that normal bone marrow cells could not reconstitute benzene-treated and lethally irradiated syngenetic recipients. As Haak[223] pointed out, this was interpreted as indirect evidence of a dearranged microenvironment in benzene toxicity. On the other hand, Kalf et al.[224] experimentally showed that benzene and its metabolites as hydroquinone and p-benzoquinone appear to poison the hemopoietic microenvironment.

The macrophage is a component of the marrow stroma, a major producer of protein factors required for maturation of marrow progenitor cells. Kalf et al.[224] studied the effects of benzene hydroquinone and p-benzoquinone on RNA synthesis in peritoneal exudate macrophage. Reincubation of peritoneal exudate macrophage with benzene, hydroquinone, and p-benzoquinone caused a dose-dependent inhibition of RNA synthesis. According to Kalf et al.,[224] benzene and its metabolites hydroquinone and p-benzoquinone may cause aplastic anemia by inhibiting the synthesis of colony-stimulating factors required for hemopoiesis. On the other hand, according to Post et al.[224a] benzene may affect hemopoiesis by damaging the bone marrow stroma that provides the microenvironment for hemopoiesis. The investigators have studied the metabolism of benzene and phenol and the effect of this chemical agent and its metabolites on macrophage RNA synthesis. Benzene is not metabolized in macrophages, but phenol is converted by peroxidase in the macrophage to both free metabolites and species which covalently bind to cellular macromolecules. Benzene and its metabolites inhibited RNA synthesis in a dose-dependent manner. According to the investigators, this inhibition was not attributable to a loss of cell viability. Benzene, possibly by an inhibition of uridine transport into macrophages, and phenol, by its conversion to covalently binding species, inhibit RNA synthesis in macrophages and thus may inhibit the synthesis of colony-stimulating factors required for hemopoiesis. Furthermore, the role of immunodeficiency in the development of benzene-mediated aplastic anemia is unknown. But as we explained earlier, there are several studies showing the decrease in T cells and changes in immunoglobulin levels in the individuals exposed to low levels of benzene and similar results in animal experiments. Therefore, it is possible that the altered immune system in benzene toxicity may play a role in the development of benzene-mediated aplastic anemia.

IV. BENZENE-INDUCED MYELOID METAPLASIA

The occurrence of myeloid metaplasia due to chronic exposure to benzene was first reported by Gall[225] in 1938. Following exposure to benzene, myeloid metaplasia developed in a worker. Gall presented detailed autopsy findings of the case. In 1941, Rawson et al.[226] claimed benzene as an etiologic factor in three workers with agnogenic myeloid metaplasia, without giving the details of the cases. In 1975, the present author and associates[134] reported myeloid metaplasia in a 35-year-old shoeworker. He was first seen 9 years ago when he was suffering from pancytopenia due to chronic exposure to benzene of 8 years duration.[37,87] At that time his bone marrow was hypercellular. After an improvement associated with discontinuance of benzene exposure, the findings of myeloid metaplasia developed. His bone marrow was hypocellular and a splenic puncture showed frank findings of myeloid metaplasia such as 20% of erythroblasts. Port et al.[227] reported a case of myeloid metaplasia due to chronic benzene exposure. We have observed two other cases of myeloid metaplasia with chronic benzene toxicity. In a 21-year-old pancytopenic technician with 4 years benzene exposure, the findings of myeloid metaplasia developed. The splenic puncture revealed the presence of extramedullary hematopoiesis. Bone marrow was very hypocellular. The third patient was a whitewasher who had the findings of myeloid metaplasia; he was exposed to benzene and also fumes of lead.[38,228]

On the other hand, surprisingly the incidence of myeloid metaplasia in chronic benzene toxicity is lower than that of leukemia, aplastic anemia, even malignant lymphoma, and multiple myeloma.[228] Our series of blood dyscrasies due to chronic exposure to benzene includes at least 47 patients with aplastic anemia, 58 with leukemia, 13 with malignant lymphoma and 4 with multiple myeloma. However, there were only three cases of myeloid metaplasia*, one following aplastic anemia.[134,228] Biscaldi and associates[229] have reported a case of transient myeloid metaplasia following an 8-year exposure to benzene. The patient showed a severe anemia associated with thrombocytopenia, reticulocytosis, and a mild bilirubinemia. Bone marrow was very hypercellular. A moderate splenomegaly appeared 1 month later, and there were immature granulocytes and erythroblasts in peripheral blood smear. Splenomegaly disappeared and bone marrow abnormalities improved 8 months later.

V. BENZENE-INDUCED PAROXYSMAL NOCTURNAL HEMOGLOBINURIA

Paroxysmal nocturnal hemoglobinuria, also known as the Machiafava-Micheli syndrome, is an extremely rare disease. It is defined by the occurrence of a population of circulating erythrocytes which are exquisitely sensitive to the hemolytic effect of complement. These red cells are assumed to be derived from one clone of abnormal stem cells.[129,231] Paroxysmal nocturnal hemoglobinuria is characterized by attacks of intravascular hemolysis and nocturnal hemoglobinuria. Yet in several cases, paroxysmal hemoglobinuria may be manifested by chronic intravascular hemolysis, without showing nocturnal character or pancytopenia. In recent years the relationship between aplastic anemia and paroxysmal nocturnal hemoglobinuria has been clearly established.[129,231] Paroxysmal nocturnal hemoglobinuria may precede or develop during the course of aplastic anemia. Some cases of paroxysmal hemoglobinuria terminates with acute myeloblastic leukemia. Even in one case of aplastic anemia, paroxysmal nocturnal hemoglobinuria and later, acute myeloblastic leukemia developed.

There are several drugs and chemicals claimed as the factors in etiology of paroxysmal nocturnal hemoglobinuria.[129,231] These are resorcin, insecticides, chloramphenicol and petrol, and a compound of a variety of chemicals and drugs such as antihistamines, antibiotics, phenacetin, etc.[129,231]

The first case of paroxysmal nocturnal hemoglobinuria associated with chronic benzene toxicity had been reported by Croizat et al.[232] The patient had worked in a tire factory and was exposed to benzene fumes. Marchal et al.[233] have reported a 42-year-old patient with paroxysmal nocturnal hemoglobinuria who worked for 2 years in a car plant with cellulose paints thought to contain benzene. In 1974, the present author and associates had reported a case of paroxysmal nocturnal hemoglobinuria in a 74-year-old male with 10 years of exposure to benzene and other solvents such as toluene in a dye factory.[134] The patient showed a moderate hepatosplenomegaly, nocturnal hemoglobinuria, findings of increased hemolysis in a mild degree, and leukopenia associated with a marked monocytosis, ranging between 24 and 36%. The test characteristics for paroxysmal nocturnal hemoglobinuria such as sucrose hemolysis and acid hemolysis were positive. Another interesting point in this patient was the marked monocytosis in the peripheral blood smear which persisted in the course of the disease. Although the degree of monocytosis varied a little in the course of

---

*   In 1947, Bowers[230] reported a case of myeloid metaplasia who had been exposed to benzene for 4 years. He had a severe refractory anemia, with hypocellular bone marrow. In his peripheral blood smear there were numerous immature elements of granulocytic series and erythroblasts ranging between 5 and 26%. In necropsy, there were numerous foci of extramedullary hematopoiesis, particularly in the liver and the spleen. Bowers was against a possible diagnosis of leukemia because the elements present in myelogram were very heterogen. Despite this, we are inclined to accept that the patient investigated by Bowers was a case of acute erythroleukemia due to chronic exposure to benzene. The predominance of young precursors of erythroid series and myeloblasts in the hypercellular bone marrow and the absence of fibrosis were evidences in this respect.

the disease, it was still impressive when he was last seen before his death. As there was no known cause for the marked monocytosis in this patient, the authors were inclined to accept it as a rare hematologic manifestation of chronic benzene toxicity. Another case of paroxysmal nocturnal hemoglobinuria due to benzene had been reported by Akman et al.,[234] because of the occurrence of intravascular coagulability during the course of illness. Some years later, we studied this patient. Although not mentioned in the paper of Akman et al.,[134] we established that the patient was a shoe maker exposed to benzene for more than 5 years.

REFERENCES

1. **Aksoy, M.,** Benzene: leukaemia and malignant lymphoma, in *Topical Reviews Haematology,* Vol. 2, Roath, S., Ed., Wright-PSG, Littleton, Mass., 1982, 105.
2. **Kluge, A.,** Properties and occurrence in nature, in *Evaluation of Benzene Toxicity in Man and Animals,* DGMK Berichte, Res. Rep. 174-6, Druckerei Hermann Lange, Hamburg, 1980, 7.
3. **Hunter, D.,** *Diseases of Occupations,* The English Universities Press, London, 1975, 471.
4. **Newsome, J. R., Norman, V., and Keith, C. H.,** Vapor phase analysis tobacco smoke, *Tob. Sci.,* 9, 102, 1965.
5. **Lawyers, R.,** Biological criteria for selected industrial toxic chemicals: a review, *Scand. J. Work Environ. Health,* 1, 139, 1975.
6. **Berlin, M.,** Low level benzene exposure in Sweden effect on blood elements and body burden of benzene, *Am. J. Ind. Med.,* 7, 365, 1985.
7. Draft Report, Technology Assessment and Economic Impact Study of an OSHA Regulation for Benzene, Arthur D. Little, Cambridge, Mass., 1977, 1.
8. Committee on Toxicology, Assembly of Life Sciences, National Research Council, *A Review of Health Effects of Benzene,* National Academy of Sciences, Washington, D.C., 1976.
9. U.S. Department of Labor, Occupational Safety and Health Administration, Occupational exposure to benzene, emergency temporary standards hearing, May 3, 1977, *Fed. Regist.,* Part IV.
10. U.S. Department of Labor, Occupational Safety and Health Administration, Occupational safety and health standards, February 10, 1978, *Fed. Regist.,* Part II.
11. **Anon.,** A new domestic poison, *Lancet,* 1, 105, 1862.
12. **Averill, C.,** Benzole Poisoning, *Br. Med. J.,* 1, 709, 1889.
13. **Santesson, G. G.,** Über Chromische Vergiftungen mit Steinkohlen Benzin. Vier Todes Falle, *Arch. Hyg.,* 31, 336, 1897.
14. Organization for Economic Cooperation and Development, Environment Directorate, Env. Air 181.04, 4th Rev. Scale D. Paris, Environment Comittee Air Management Policy Group, drafted December 23, 1983, 45.
15. **Gerarde, H. W.,** *Toxicology and Biochemistry of Aromatic Hydrocarbons, Elsevier monographs on toxic agents,* Browning, E., Ed., Elsevier, Amsterdam, 1960, 98.
16. **Hayden, J. W. and Comstock, E. G.,** The clinical toxicology of solvent abuse, *Clin. Toxicol.,* 9, 169, 1976.
17. **Winek, C. H. and Collum, W. D.,** Benzene and toluene fatalities, *J. Occup. Med.,* 13, 259, 1971.
18. **Moeschlin, S.,** *Klinik und Therapie der Vergiftüngen,* Georg Thieme Verlag, Stuttgart, 1952, 246.
19. **Browning, E.,** *Toxicity and Metabolism of Industrial Solvent,* Amsterdam, Elsevier, 1965, 3.
20. **Täuber, J. B.,** Instant benzol death, *J. Occup. Med.,* 12, 91, 1970.
21. **Derot, M. and Philbert, M.,** Aique par ingestion de benzene: syndrome hepatorenal aique, *Bull. Mem. Hop. Paris,* 72, 103, 1956.
22. **Holliday, M. C., Engelhardt, F. R., Henderson, J. F., and Holmes, A.,** Occupational Health Implications of Benzene in Ontario, Standard and Programs Branch, Ontario Ministry of Labor, December 1979.
23. DFG (Deutsche Forschungsgemeinschaft), *Benzene in the Working Environment,* Harald Boldt Verlag, Boppard, W. Germany, 1974.
24. The autopsy report of Forensic Institute of Istanbul, affiliated with Cerrahpaşa Medical Faculty, 1984.
25. U.S. Department of Labor, Occupational Safety and Health Administration, Occupational exposure to benzene, proposed standards hearing, May 27, 1977, *Fed. Regist.,* Part VI, p. 27463.
26. **Infante, P. P. and White, M. C.,** Benzene: epidemiologic observations of leukemia by cell type and adverse health effects associated with low-level exposure, *Environ. Health Perspect.,* 52, 75, 1983.

27. **Selling, L.**, Preliminary report of some cases of purpura hemorrhagica due to benzene poisoning, *Bull. Johns Hopkins Hosp.*, 21, 33, 1916.

28. **Selling, L.**, Benzol a leucotoxine. Studies on the degeneration of the blood and hematopoietic organs, *Johns Hopkins Hosp. Rep.*, 17, 83, 1916.

29. **Greenburg, L.**, Benzol Poisoning. As an industrial hazard. U.S. Treasury Department Public Health Reports, Vol. 41, Part 2, Numbers 27—53, July—December 1926, and 1410—1432, 1516—1539, 1927, Public Health Service, U.S. Treasury Department, Washington, D.C.

30. **Koranyi, B. A.**, *17th Int. Congr. Med.*, part 2, Sec. 6, Oxford University Press, London, 1913; as cited in **Browning, E.**, *Toxicity and Metabolism of Industrial Solvents*, Elsevier, Amsterdam, 1965, 26.

31. **Rochner, F. J., Boldrige, C. W., Hausmann, G. H.**, Chronic benzol poisoning, *Proc. Soc. Exp. Biol. Med.*, 33, 323, 1925.

32. **Schneider, H.**, Zur Klinik und Therapie der chronischen gewerblichen ''Benzolvergiftung'', *Med. Klin. (Munich)*, 26, 1110, 1930.

33. **Hamilton, A.**, Benzene (benzol) poisoning. General review, *Arch. Pathol. Lab. Med.*, 11, 434, 1931.

34. **Greenburg, L., Meyers, R., Goldwater, L., Smith, H. R.**, Benzene poisoning in the retrogravure industry in New York City, *Ind. Med. Toxicol.*, 21, 395, 1939.

35. **Frank, E.**, Blood picture in benzene poisoning, in *Klinik Dersleri* (Turkish), Rössler, R., Ed., Ege Matbaasi, Istanbul, 1951, 278.

36. **Aksoy, M.**, Leukemia in workers due to occupational exposure to benzene. *New Istanbul Contrib. Clin. Sci.*, 12, 3, 1977.

37. **Aksoy, M. and Erdem, Ş.**, Follow-up study on the mortality and the development of leukemia in 44 pancytopenic patients with chronic exposure to benzene, *Blood*, 52, 285, 1978.

37b. **Tangün, Y., Öngür, E., Inceman, Ş.**, Son aylarda deri-iş endüstri kollarinda çalişanlarda gördüğümüz benzol entoksikasyonuna bağğli 6 aplastik anemi vakasi, *Bull. Turkish Med. Soc.*, 28, 33, 1967 (Turkish).

38. **Aksoy, M.**, Different types of malignancies due to occupational exposure to benzene: a review of recent observations in Turkey, *Environ. Res.*, 23, 181, 1980.

39. **Goldstein, B. D.**, Hematotoxicity on humans, in *Benzene Toxicity, A Critical Evaluation*, Laskin, S. and Goldstein, B. D., Eds., McGraw-Hill, New York, 1978; *J. Toxicol. Environ. Health*, Suppl. 2, 69, 1977.

40. **Savilahti, M.**, Mehr als 100 Vergiftungsfälle durch Benzol in einer Schuch Fabrik, *Arch. Gewerbepathol. Gewerbehyg.*, 15, 147, 1956.

41. **Lob, M.**, Role etiologie du benzenique dans deux hemopathies mortelles pretandeux cryptogenique. *Schweiz. Med. Wochenschr.*, 99, 1181, 1969.

42. **Hirokowa, M. T.**, Some observations on chronic benzene intoxication, *Arch. Mal. Prof. Hyg. Toxicol. Ind.*, 21, 46, 1960.

43. **Böttiger, L. E.**, Epidemiology of aplastic anemia, in *Aplastic Anemia, Pathophysiology and Approaches to Therapy*, Heimpel, H., Gordon-Smith, E. C., Heit, W., and Kubaneck, B., Eds., Springer-Verlag, Basel, 1979, 27.

44. **Sakol, M. J.**, Testimony at OSHA hearing on benzene, U.S. Occupational Safety and Health Administration, Washington, D.C., 1977.

45. **Yin, S.-N., Li, Q., Liu, Y., Tian, F., and Jin, C.**, Study on benzene toxicity, Symp. Inst. Health, People's Republic of China, 1983.

45a. **Yin, S.-N., Li, Q., Tian, F., Du, C., and Jin, C.**, Occupational exposure to benzene in China, *Br. J. Ind. Med.*, 44, 192, 1987.

46. **Brandt, L., Nilsson, P. G., and Mitelman, F.**, Non-industrial exposure to benzene as leukemogenic risk factor, *Lancet*, 2, 1047, 1977.

47. **Aksoy, M., Erdem, Ş., Dinçol, G., Bakioğlu, I., and Kutlar, A.**, Aplastic anemia due to chemicals and drugs: a study of 108 patients, *Sex. Transmit. Dis.*, Suppl. 11, 347, 1984.

48. **Reifschneider, C. A.**, Benzol: its occurrence and prevention, Natl. Safety Council. 11th Annu. Congr. Proc. Detroit, 1922, 249; as cited in **Browning, E.**, *Toxicity and Metabolism of Industrial Solvents*, Elsevier, Amsterdam, 1965.

49. **Ronchetti, V.**, *Atti Soc. Lomb. Sci. Med. Biol.*, 2, 322, 1922; as cited in **Hunter, D.**, *Diseases of Occupations*, English Universities Press, London, 1975, 483.

50. **Aksoy, M. and Erdem, Ş.**, Some problems of hemoglobin patterns in different thalassemic syndromes showing the heterogeneity of beta-thalassemia genes, *Ann. N.Y. Acad. Sci.*, 165, 13, 1969.

51. **Aksoy, M., Erdem, Ş., Erdoğan, G., and Dinçol, G.**, Acute leukemia in two generations following chronic exposure to benzene, *Hum. Hered.*, 24, 70, 1974.

52. **Aksoy, M.**, Unpublished data.

53. **Baslo, A. and Aksoy, M.**, Neurological abnormalities in chronic benzene poisoning. A study of six patients with aplastic anemia and two with preleukemia, *Environ. Res.*, 27, 457, 1982.

54. **Aksoy, M.**, Malignancies due to occupational exposure to benzene, *Am. J. Ind. Med.*, 7, 395, 1985.

55. **Goldstein, B. D.**, Clinical hematotoxicity of benzene, in *Advances in Modern Environmental Toxicology: Carcinogenecity and Toxicity of Benzene*, Vol. 4, Mehlman, M. A., Ed., Princeton Scientific Publishers, Princeton, N.J., 1983, 51.

56. **Hirokowa, M. T.**, Quelques observationes sur l'intoxication benzenique, *Arch. Mal. Prof. Hyg. Toxicol. Ind.*, 21, 46, 1960.

57. **Ito, T.**, Study on sex difference in benzene poisoning. Report on the obstacles in benzene workers, *Igakukai Zarahi*, 22, 263, 1962; as cited in **Browning, E.**, *Toxicity and Metabolism of Industrial Solvents*, Elsevier, Amsterdam, 1965.

58. **Siou, C., Conan, L., and Haitem, M. El.**, Evaluation of the clastogenic action of benzene by oral administration with cytogenic techniques in mouse and chinese hamster, *Mutat. Res.*, 90, 273, 1981.

59. **Gad-El-Karim, M. M., Harper, B. L., and Legator, M. S.**, Modifications in the myelocytogenic effect of benzene in mice with toluene, Phenobarbital, 3, methylcholanthrene, Arocolor, 1254 and SKF, *Mutat. Res.*, 195, 225.

60. **Gad-El-Karim, M. M., Ramunojam, V. M. S., Ahmed, E. A., and Legator, M. S.**, Benzene myeloclastogenicity: a function of its metabolism, *Am. J. Ind. Med.*, 7, 475, 1985.

61. **Sato, A., Nakijama, T., Fujiware, Y., and Murayama, N.**, Kinetic studies on sex difference in susceptibility to chronic benzene intoxication with special reference to body fat content, *Br. J. Ind. Med.*, 32, 321, 1975.

62. International Labour Office, Geneva, Benzene Report of the Meeting of Experts, 1968, 4.

63. **Doskin, T. A.**, Effect of age on the reaction to a combination of hydrocarbons, *Hyg. Sanit.*, 36, 379, 1971.

64. **Aksoy, M., Dinçol, K., Akgün, T., Erdem, Ş., and Dinçol, G.**, Haematological effects of chronic benzene poisoning in 217 workers, *Br. J. Ind. Med.*, 28, 296, 1971.

65. **Aksoy, M., Erdem, Ş., and Dinçol, G.**, Types of leukemia in chronic benzene poisoning. A study in thirty-four patients, *Acta Haematol.*, 55, 65, 1976.

66. **Weaver, N. K., Gibson, R. L., and Smith, L. W.**, Occupational exposure to benzene in the petroleum and petrochemical industries, in *Advances in Modern Environmental Toxicology: Carcinogenecity and Toxicity of Benzene*, Vol. 4, Mehlman, M. A., Ed., Princeton Scientific Publishers, Princeton, N.J., 1983, 63.

67. **Aksoy, M.**, Testimony to Information Hearing on the Proposed OSHA Benzene Standard, Washington, D.C., July 20, 1977.

68. **Aksoy, M.**, Benzene and leukemia, *Lancet*, 1, 441, 1978.

69. **Aksoy, M.**, Benzene leukemogenic effects and exposure limits, in 43 ILO Symp. on New Trends in the Optimisation of the Working Environment, Istanbul, Occupational Safety and Health Series, Geneva, 1979, 336.

70. **Zenz, C.**, Benzene attempts to establish a lower exposure standard in the United States: a review, *Scand. J. Work Environ. Health*, 4, 103, 1978.

71. News of the week, High Court overturns OSHA benzene rule, July 7, 1980, 4, C8.

72. **Fishbein, G. W.**, The Supreme Court's benzene strikes at basic public health principles, *Occup. Health Safety Lett.*, 10, 1, 1980.

73. **Townsend, J. C., Ott, M. G., and Fishbeck, W. A.**, Health exam findings among individuals occupationally exposed to benzene, *J. Occup. Med.*, 20, 543, 1978.

74. **Tsai, S. P., Wen, C. P., Weiss, N. S., Wong, O., McClellan, W. A., and Gibson, R. L.**, Retrospective mortality and medical surveillance studies of workers in benzene areas of refineries, *J. Occup. Med.*, 25, 685, 1983.

75. **Kahn, H. and Muzyka, V.**, The chronic effect of benzene on porphyrine metabolism, *Work Environ. Health*, 110, 140, 1973.

76. **Girard, R., Mallein, M. L., Bertholon, J., Coeur, P., and Tolot, F.**, Leucocyte alcaline phosphatase et exposes du benzene, *Med. Lav.*, 61, 50, 1970.

77. **Though, I. M. and Court-Brown, W. M.**, Chromosome observations and exposure to ambient benzene, *Lancet*, 1, 684, 1965.

78. **Picciani, D.**, Cytogenetic study of workers exposed to benzene, *Environ. Res.*, 19, 33, 1979.

79. **Ott, M. G., Townsend, J. C., Fishbeck, W. A., and Lagner, R. H.**, Mortality among individuals occupationally exposed to benzene, *Arch. Environ. Health*, 33, 3, 1978.

80. **Baarson, K. A., Snyder, C. A., and Albert, R. E.**, Repeated exposure of C57B1 mice to inhaled benzene at 10 ppm markedly depressed erythropoietic colony formation, *Toxicol. Lett.*, 20, 337, 1984.

81. **Chang, I. W.**, Study on threshold limit value of benzene and early diagnosis of benzene poisoning, *J. Cat. Med. Coll.*, 23, 429, 1972.

82. **Truhaut, R.**, Discussions on permissible limit for benzene exposure, Meet. of the Royal Soc. of Medicine, London, August 30—31, 1983.

83. **Holmberg, B. and Lunberg, P.**, Benzene standards, occurrence and exposure, *Am. J. Ind. Med.*, 7, 375, 1985.

83a. **Aksoy, M.**, Malignancies and occupational exposure to benzene, in *Homage Au Professeur Rene Truhaut*, Paris, 1984, 4.

84. **Lachnit, V. and Reimer, E. E.**, Panmyelopathien durch Aromatische Lösungmittel, *Wien. Klin. Wochenschr.*, 71, 365, 1959.

85. **Saita, G. and Perini, A.**, Dos cas da benzolisme nel lavoro in casa, *Med. Lav.*, 49, 442, 1958.

86. **Dési, I.**, *Int. Arch. Gewerbepathol. und Gewerbehyg.*, 23, 58, 1967; as cited in DFG (Deutsche Forschungsgemeinschaft), *Benzene in the Work Environment*, Harald Boldt Verlag, Boppard, W. Germany, 1974.

87. **Aksoy, M., Dinçol, K., Erdem, Ş., Akgün, G., and Dinçol, G.**, Details of blood changes in 32 patients with pancytopenia associated with long-term exposure to benzene, *Br. Ind. Med.*, 29, 56, 1972.

88. **Koslova, T. A.**, The effect of benzene on the organism at high air temperature, *Gig. Sanit.*, 22(Abstr.), 18, 1957.

89. **Hill, B. W. and Venable, L.**, The interaction of ethylalcohol and industrial chemicals, *Am. J. Ind. Med.*, 3, 321, 1982.

90. **Aksoy, M.**, Unpublished data.

91. **Snyder, C. A., Baarson, K. A., Goldstein, B. D., and Albert, R. E.**, Ingestion of ethanol increases the hematotoxicity of inhaled benzene C57B1 mice, *Bull. Environ. Contam. Toxicol.*, 27, 175, 1981.

92. **Baarson, K. A., Snyder, C. A., Green, J. D., Sellakumar, A., Goldstein, B. D., and Albert, R. E.**, Hematotoxic effects of inhaled benzene on peripheral blood bone marrow and spleen cells are increased by ingested ethanol, *Toxicol. Appl. Pharmacol.*, 64, 393, 1982.

93. **Rosenthal, G. J. and Snyder, C. A.**, The effects of ethanol and the role of the spleen during benzene-induced hematotoxicity, *Toxicology*, 30, 283, 1984.

94. **Driscoll, K. E. and Snyder, C. A.**, The effects of ethanol ingestion and repeated benzene exposures on benzene pharmacokinetics, *Toxicol. Appl. Pharmacol.*, 73, 525, 1984.

95. **Saita, G. and Moreo, L.**, Talassemio e benzolisms cronico, *Med. Lav.*, 50, 25, 1959.

96. **Gaultier, P. M., De Traverse, P. M., Coquelet, M. L., Loygue, A. M., Housset, H., and Gervais, P.**, Hemoglobinopathies observes in milieu professionel, *Bull de l'INSERMT*, 21, 1047, 1966.

97. **Saita, G. and Moreo, L.**, Crisi emolitica da inolazione di benzole in dose ünica massiva, *Med. Lav.*, 52, 713, 1961.

98. **Simson, R. E. and Shandlar, A.**, Thalassemia minor in industry, *Med. J. Aust.*, 801, 1971.

99. **Monaenkova, A. M. and Zorina, L. A.**, Hemodynamics and heart muscle changes in chronic benzene poisoning, *Gig. Tr. Prof. Zabol.*, 4, 30, 1975 (Russian); *Ind. Hyg. Dig.*, (Abstr.) 1035, 1975.

100. **Godrot, J., Quercy, J., and Pedan, P.**, Hemopathie atypique consecutive a une interaction benzolique chroniques, *Arch. Mal. Prof. Hyg. Toxicol. Ind.*, 20, 121, 1959.

101. **Mallory, T. B., Gall, E. A., and Brickley, W. J.**, Chronic exposure to benzene, *J. Ind. Hyg. Toxicol.*, 21, 355, 1939.

102. **Goldwater, L. J.**, Disturbances in the blood following exposure to benzol, *J. Lab. Clin. Med.*, 26, 957, 1941.

103. **Hamilton-Paterson, J., Browning, E.**, Toxic effects in woman exposed to industrial rubber solutions, *Br. Med. J.*, 1, 340, 1944.

104. **Danysz, M.**, Quelques resulates hematologiques constates chez des ouveriers travaillant dans la benzene, *Sang*, 15, 348, 1942.

105. **Bernard-Pichon, A.**, Incertitu des at difficultes de la preventation du benzolism par la surveillance hematologique systematique, *Sang*, 15, 340, 1942.

106. **Paterni, L. and Russo, G.**, Ilprimo casi'di anomalia pelgherina nella emopatia benzenica, associazone con manismo cellulare, *Haematologica*, 45, 213, 1960.

107. **Saita, G. and Moreo, L. A.**, A case of chronic benzene poisoning with a Pelger-Huet type leucocyte anomaly, *Med. Lav.*, 52, 331, 1966.

108. **Zini, C. and Alessandri, M.**, Anomalia leukociteria pseudo pelgeriane in un caso di emopatia benzonica can leucosi acuta terminalle, *Haematologica*, 52, 258, 1967.

109. **Latta, J. S. and Davies, L. T.**, Effects of the blood and hemopoietic organs of the albino rat of repeated administration of benzene, *Arch. Pathol.*, 31, 55, 1941.

110. **Irons, R. D., Wierda, D., and Pfeifer, R. W.**, The immunotoxicity of benzene and its metabolites, in *Advances in Modern Environmental Toxicology, Carcinogenecity and Toxicity of Benzene*, Vol. 4, Mehlman, M. A., Ed., Princeton Scientific Publishers, Princeton, N.J., 1983, 37.

111. **Greenlee, W. F. and Irons, R. D.**, Modulation of benzene induced lymphocytopenia in the rat by 2,4,5,2',4',5'-hexachlorobiphenyl and 3,4,3',4'-tetrachlorobiphenyl, *Chem. Biol. Interact.*, 33, 345, 1981.

112. **Moszczynski, P. and Lisiewicz, J.**, Occupational exposure to benzene, toluene and xylene and T lymphocyte functions, *Haematologica*, 17, 449, 1984.

113. **Moszczynski, P. and Lisiewicz, J.**, Hematological indices of peripheral blood in workers occupationally exposed to benzene, toluene and xylene, *Zb. Bakt. Hyg.*, 178, 329, 1984.

114. **Moszczynski, P. and Lisiewicz, J.,** Occupational exposure to benzene, toluene and xylene and the lymphocyte lysosomal N-acetyl-beta-D-glucosaminidase, *Ind. Health,* 23, 47, 1985.

115. **Bernard, J.,** La lymphocytose benzenique, *Sang,* 15, 501, 1942.

116. **Girard, R., Mallein, M. L., Bertholon, J., and Coeur, P. J. Cl.,** Étude de la phosphatase alcaline leucocyteire et du caryotype des ouvriers exposés au benzene, *Arch. Mal. Prof. Med. Trav. Secur. Soc.,* 31, 31, 1970.

117. **Songnian, Y., Quilan, L., and Yuxuang, L.,** Significance of leukocyte alkaline phosphatase in diagnosis of chronic benzene poisoning, *Reg. Toxicol. Pharmacol.,* 2, 209, 1982.

118. **Pollini, G. and Colombi, R.,** Modificazioni delle resistenze osmotiche leucocitorie nei soggetti a rischio benzolico *Lav. Um.,* 16(1), 177, 1964.

119. **Volkova, A. P.,** Action of benzene intoxication on the phagocytic activity of rabbit leucocytes *Gig. Sanit.,* 24, 80, 1959; *J. Med.,* 2(Abstr.), 195, 1960.

120. **Koslova, T. A. and Volkova, A. P.,** The blood picture and phagocytic activity of leucocyte in men having contact with benzol, *Gig. Sanit.,* 25, 29, 1960.

121. **Boiko, V. I., Makarieva, L. M.,** Conditions of work and health status of workers engaged in production of benzene and its homologs from crude oil, *Gig. Tr.; Ind. Hyg. Dig.,* 42(Abstr.), 25, 1978.

122. **Moszczynski, P.,** The effect of work environment contaminated with organic solvents on the enzymatic composition of neutrophils, *Haematologica,* 67, 482, 1982.

123. **Moszczynski, P. and Lisiewicz, P.,** Effect of environmental contamination of the work place with benzene, toluene and xylene on human lymphocyte associated immunity, *Med. Lav.,* 74, 792, 1983.

124. **Lange, A., Smolik, R., Zatonski, W., and Glazman, H.,** Leukocyte agglutinins in workers exposed to benzene, toluene and xylene, *Int. Arch. Arbeitsmed.,* 31, 45, 1973.

125. **Heim de Balzak, and Lafont, A. E.,** Reactions hematique du benzenique, Prof. Assoc. France pour Avancement de Science, Congr. de Toulore; as cited in **Browning, E.,** *Toxicity and Metabolism of Industrial Solvents,* Elsevier, Amsterdam, 1965.

126. **Duvoire, M. and Derobert, L.,** L'eosinophile de benzenique, *Arch. Mal. Prof. Hyg. Toxicol. Ind.,* 15, 241, 1942.

127. **Hernberg, S., Savilahti, M., Ahlman, K., and Asp, S.,** Prognostic aspects of benzene poisoning, *Br. J. Ind. Med.,* 23, 204, 1966.

128. **Sungur, T.,** *Intoxication of Benzene* (Turkish), Başnur Printing House, Ankara, 1969.

129. **Winetrobe, M. M., Lee, G. R., Boggs, D. R., Bethell, T. C., Athens, J. W., and Forester, J.,** *Clinical Hematology,* 7th ed., Lea & Febiger, Philadelphia, 1974, and 8th ed., 1981, 700.

130. **Smith, A. R.,** Chronic benzene poisoning among woman industrial workers, *J. Ind. Hyg. Toxicol.,* 10, 73, 1928.

131. **Meşulem, V.,** The role of absolute basophil counting in various hematological disorders, No. 86, ISH European-African Div. 4th Meet. Abstracts, Lectures and Symposia, September 5—9, 1977.

132. **Aksoy, M., Özeriş, S., Sabuncu, H., Inanici, Y., and Yanardağ, R.,** A haematological study on 231 workers exposed to benzene in Istanbul and Izmit, during the period of 1983 and 1985, *Br. J. Ind. Med.,* in press.

133. **Roth, L., Turcanu, P., Dinu, I., and Moise, G.,** Monocytosis in those who work with benzene and chronic benzene poisoning, *Folia Haematol.,* 100, 213, 1973.

134. **Aksoy, M., Erdem, Ş., and Dinçol, G.,** Two rare complications of chronic benzene poisoning: myeloid metaplasia and PNH. Report of 2 cases, *Blut,* 255, 1975.

135. **Grazioli, G., Monteverdi, C. A.,** Survey of Italian literature on industrial medicine, *Panminerva Med.,* 2, 142, 1960.

136. **Solov'eva, E. A.,** Morphological changes in blood platelets and of megakaryocytes in chronic benzene poisoning, *Gig. Tr. Prof. Zabol.,* 7, 57, 1963; as cited in **Laskin, S. and Goldstein, B. D.,** *J. Toxicol. Environ. Health,* Suppl. 2, 139, 1977.

137. **Craveri, A.,** La fibrinolisi le piastrine le fibrinogeno avri-tests emocoagulativinel benzolismo clinico, *Med. Lav.,* 53, 722, 1962.

138. **Saita, G., Sbertoli, C., and Parina, G. F.,** Indagini trombo elastografiche Nell'emopatia benzolica, *Med. Lav.,* 55, 655, 1964.

139. **Kuhbeck, J. and Lachnit, V.,** Zum Problem der Früherfassung von Benzol und Toluolschöden, *Int. Arch. Gewerbepathol. Gewerbehyg.,* 19, 149, 1962.

140. **Hunter, F. T.,** Chronic exposure to benzene (benzol), *J. Ind. Hyg. Toxicol.,* 21, 331, 1939.

141. **Villiani, C., Massei, C., and Zanobini, R.,** Erythropoiesis of experimental benzene hemopathy, *Folia Med. (Naples),* 43, 842, 1962.

142. **Truhaut, R., Paoletti, C., Boiron, M., and Tubiana, M.,** Etudes experimentales sur la toxicologie du benzene. Modifications del'erythropoiese survies par le fer radioactif, *Arch. Mal. Prof. Hyg. Toxicol. Ind.,* 20, 9, 1959.

143. **Snyder, R. and Kocsis, J. J.,** Current Concepts of Chronic Benzene Toxicity, *Crit. Rev. Toxicol.,* 3, 265, 1975.

144. **Paolino, W., Resegotti, L., and Sartoris, S.,** Compertamano dei sulfidril serici el tessutalinella intossicazione sperment e de benzolo, *Folia. Med. (Naples),* 43, 763, 1961.

145. **Biancacchio, A. and Fermariello, V.,** Protoporfirine libere eritrocitoire e coproporfirino urinaire nel'outossicazione subcronica sperimentale da benzol, *Folia. Med. (Naples),* 43, 588, 1961.

146. **Kahn, H. and Muzyka, V.,** The chronic effect of benzene on porphyrine metabolism, *Work Environ. Health,* 10, 40, 1973.

147. **Zatonski, W.,** Studies of activity of glycolitic enzymes, with special reference to phosphofructokinase properties, in erythrocytes of persons exposed to benzene, *Prace Naukowe AM,* 8, 83, 1974 (English abstract).

148. **Hancke, J. Z.,** Preliminary investigations of prolonged occupational exposure to toxic substances on the level of some serum enzymes, *Arch. Hig. Rad.,* 15, 57, 1964.

149. **Lange, A., Smolik, R., Zatonski, W., and Szymanska, J.,** Serum immunoglobulin levels in workers exposed to benzene, toluene and xylene, *Int. Arch. Arbeitsmed.,* 31, 37, 1973.

150. **Smolik, R., Grzybek-Hryncewicz, A., Lange, A., and Zatonski, W.,** Serum complement level in workers exposed to benzene, toluene and xylene, *Int. Arch. Arbeitsmed.,* 31, 233, 1973.

151. **Sobczyk, W., Siedlecka, B., Gajewska, Z., Horyol, W., and de Mezer, O.,** EEG recordings in workers exposed to benzene compounds, *Med. Pr.,* 24, 273, 1973 (Polish, English abstract).

152. National Institute for Occupational Safety and Health, Revised Recommendation for an Occupational Exposure Standard for Benzene, Center for Disease Control, Public Health Service, Department of Health, Education and Welfare, Atlanta, 1976.

153. **Hanke, J. T., Dutakiewics, R., and Piotrowski, J.,** The absorption of benzene through skin in men, *Med. Pr.,* 12, 413, 1961; as cited in **Täuber, U.,** in *Evaluation of Benzene Toxicity in Man and Animals,* DGMK Berichte, Res. Rep. 174-6, Druckerei Hermann Lange, Hamburg, 1980, 25.

154. **Maibach, H. I.,** Percutaneous penetration of benzene: Man, in letter from Rathbum, D. B., American Petroleum Institute to Auchtes, T. G., Department of Labor, March 1981; as cited in **Infante, P. P. and White, M. C.,** *Environ. Health Perspect.,* 52, 75, 1983.

155. **Wester, R. E. and Maibach, H. I.,** Percutaneous absorption. A perspective, in *Cutaneous Toxicity,* Drill, V. and Lazar, P., Eds., Academic Press, New York, 1977, 240; as cited in **Täuber, U.,** *Evaluation of Benzene Toxicity in Man and Animals,* DGMK Berichte, Res. Rep. 174-6, Druckerei Hermann Lange, Hamburg, 1980, 25.

156. U.S. Department of Labor, Occupational Safety and Health Administration, Occupational exposure to benzene, emergency temporary standards hearing, May 3, 1977, *Fed. Regist.,* Part IV, p. 22517.

157. **Täuber, U.,** Absorption, metabolism, distribution, elimination and detection, in Evaluation of Benzene Toxicity in Man and Animals, DGMK Berichte, Res. Rep. 174-6, Druckerei Hermann Lange, Hamburg, 1980, 25.

158. **Susten, A. S., Dames, B. L., Burg, J. R., and Niemeier, R. W.,** Percutaneous penetration of benzene in hairless mice an estimate of dermal absorption during tire-building operations, *Am. J. Ind. Med.,* 7, 323, 1985.

159. **Alter, B. P., Potter, H. U., and Li, P. P.,** Classification and etiology of the aplastic anemias, *Clin. Haematol.,* 7, 431, 1978.

160. **Smick, K. M., Condit, P. K., Proctor, R. L., and Sutcher, V.,** Fatal aplastic anemia. An epidemiological study of its relationship to the anemia drug chloramphenicol, *J. Chronic Dis.,* 17, 899, 1964.

161. **Wallerstein, R. O., Condit, P. K., Kasper, C. K., Brown, J. W., and Morrison, F. R.,** Statewide study of chloramphenicol therapy and fatal aplastic anemia, *JAMA,* 208, 2045, 1969.

162. **Gale, P. R., Champlin, R. E., Feig, S. A., and Fitchen, J. H.,** Aplastic anemia: biology and treatment, *Ann. Intern. Med.,* 95, 477, 1981.

163. **Heimpel, H.,** Aplastic anemia, in Trends in Haematology, Part I, 8th Meet. of European-African Div. of ISH, September 8—13, 1985, Warsaw, Poland, 181.

164. **Whang, K. S.,** Aplastic anemia in Korea: a clinical study of 309 cases, in *Aplastic Anemia,* Proc. of the 1st Int. Symp. on Aplastic Anemia, Japan Medical Research Foundation Ed., University of Tokyo Press, Tokyo, 1978, 225.

165. **Erf, L. A. and Rhodas, C. P.,** The hematological effects of benzene (benzol) poisoning, *J. Ind. Hyg. Toxicol.,* 21, 421, 1939.

166. **Gorini, P., Colombi, R., and Peconati, O.,** Investigation on the origin of the macrocytosis in the cytoplasmic anemia from chronic benzene poisoning, *Lav. Um.,* 11, 121, 1959.

167. **Curletto, R. and Ciconali, M.,** Haematological disorders in benzene poisoning, *Med. Lav.,* 53, 505, 1962.

168. **Corsico, R., Biscaldi, G. P., and Lalli, M.,** Benzene induced changes in the size of hemopoietic and haemic cells, *Lav. Um.,* 19, 16, 1967.

169. **Helmer, J.,** Et 60-tal fallav kronisk Benzol-forgiftning vid Gummifabrik, *Nord. Med.,* 21, 288, 1944; as cited in **Savilahti, M.,** *Arch. Gewerbepathol. Gewerbehyg.,* 15, 147, 1956.

170. **Barret, P. V. D., Cline, M. J., and Berlin, N. I.**, The association of the urobilin "early peak" and erythropoiesis in man, *J. Clin. Invest.*, 45, 1657, 1966.
171. **Dacie, J. V. and Lewis, S. M.**, *Practical Hematology*, 2nd ed., Blackwell Scientific, Oxford, 1963.
172. **Aksoy, M., Erdem, Ş., Akgün, T., Okur, Ö., and Dinçol, K.**, Osmotic fragility studies in three patients with aplastic anemia due to chronic benzene poisoning, *Blut*, 13, 85, 1966.
173. **Kinkead, E. R., Di Pasquale, L. O., Vernet, E. H., and MacEwen, J. D.**, Chronic Toxicity of JP-4 Jet Fuel, NTS-Aerospace Med. Res. Lab., Wright-Patterson Air Force Base, Ohio, 1974.
174. **Paolino, W., Reseqetti, L., and Sartoris, S.**, Compertemano dei sulfidril serici e tessutalinella intossicazione spermentole de benzols *Folia Med. (Naples)*, 43, 763, 1961; as cited in **Browning, E.**, *Toxicity and Metabolism of Industrial Solvents*, Elsevier, Amsterdam, 1965.
175. **Hektoen, L.**, Effect of benzene on production of antibodies, *J. Infect. Dis.*, 19, 69, 1916; as cited in **Browning, E.**, *Toxicity and Metabolism of Industrial Solvents*, Elsevier, Amsterdam, 1965, 3.
176. **Aksoy, M. and Seçer, F.**, Fetal hemoglobin in acquired aplastic anemia, *Acta Haematol.*, 32, 188, 1964.
177. **Aksoy, M., Erdem, Ş., Schroeder, W. A., and Huisman, T. H. J.**, Hemoglobins A_2 and F in chronic benzene poisoning, in *Int. Istanbul Symp. on Abnormal Hemoglobins and Thalassemia*, Aksoy, M., Ed., TBTAK (Scientific and Technical Research Council of Turkey), Angora 1975, 197.
178. **Bloom, C. E. and Diamond, L.**, Prognostic value of fetal hemoglobin levels in acquired aplastic anemia, *N. Engl. J. Med.*, 278, 304, 1968.
179. **Aksoy, M., Erdem, Ş., and Dinçol, K.**, Fetal hemoglobin, correspondence, *Blood*, 41, 742, 1973.
180. **Storti, E., Perugini, S., Mussini, C., and Manzini, E.**, Compertamento della emoglobina fetale nelle mielopatic involutive e suo interesse prognistica, in *Le Mielopatié Involutive*, Storti, E. and Perugini, S., Eds., Pozzi, Rome, 1969, 426.
181. **Aksoy, M., Dinçol, K., Erdem, Ş., and Akgün, T.**, The result obtained from a triple treatment with androgens, steroids and phytohemogglutinine of aplastic anemia of various etiology with emphasis on the prognostic value of of foetal haemoglobin content and bone marrow cellularity, 5th Congr. Asian Pacific Soc. Haematology Abstract Book, September 1—6, 1969, Istanbul, 31.
182. **Storti, E., Perugini, S., Mussumi, S., and Manzini, E.**, Comportamento della emoglobina fetale nelle mielopatic involutive suo interesse prognistico, in *Le Mielopatié Involutive*, Storti, E. and Perugini, S., Eds., Pozzi, Rome, 1969, 426.
183. **Shahidi, N. and Diamond, K.**, Alkali resistent hemoglobin in aplastic anemia of both congenital and acquired type, *N. Engl. J. Med.*, 266, 117, 1962.
184. **Aksoy, M., Erdem, Ş., and Dinçol, G.**, The reaction of normal and thalassemic individuals to benzene poisoning: the diagnostic significance of such studies, in *Abnormal Haemoglobins and Thalassemia. Diagnostic Aspects*, Schmidt, R. M., Ed., Academic Press, New York, 1975, 267.
185. **Aksoy, M. and Erdem, Ş.**, Decrease in the concentrations of HbA_2 during the course of erythroleukemia, *Nature (London)*, 213, 522, 1967.
186. **Weatherall, D. J., Edwards, J. A., and Donohoe, W. T. A.**, Haemoglobin and red cell enzyme changes in Juvenile myeloid leukaemia, *Br. Med. J.*, 1, 679, 1968.
187. **Aksoy, M., Erdem, Ş., Dinçol, K., and Akgün, T.**, Haemothorax and haemorrhagic pleurisy as rare and severe complications of aplastic anemia. Report of 2 cases, *Punjab. Med. J.*, 19, 42, 1969.
187a. **De Gruchy, G. C.**, *Drug-Induced Blood Disorders*, Blackwell Scientific, Oxford, 1975, 46.
187b. **Aksoy, M., Keskin, H., Erdem, Ş., and Dinçol, G.**, A new syndrome: benzene-hepatitis-aplastic anemia syndrome. Report of a case, paper presented at the Meeting of the Turkish Society of Hematology, Bursa, Turkey, February 23 to 25, 1987, will be published.
188. **Wirtschafter, Z. T. and Cronyn, M. W.**, Relative hepatotoxicity, *Arch. Environ. Health*, 9, 180, 1962.
189. **Rozera, G., Elmino, O., and Colicchio, G.**, Hematic studies, blood coagulation, liver function in benzene experiments, *Folia Med. (Naples)*, 42, 1373, 1960.
190. **Truhaut, R.**, Reporter, *Benzene: Uses, Toxic Effects, Substitutes*, Meeting of Experts on the Safe Use of Benzene and Solvents Containing Benzene, International Labor Office, Geneva, 1968, 1.
191. **Ikeda, M.**, Poisoning due to benzene and its substitutes in Japan, paper presented at the Proc. Int. Workshop of Toxicology of Benzene, Paris, November 9—11, 1976, 1.
192. **Herskowitz, A., Ishü, N., and Schaumberg, H.**, n-Hexane neuropathy. A syndrome occurring as a result of industrial exposure, *New Engl. J. Med.*, 285, 82, 1971.
193. **Matsumato, T.**, Experimental studies on the chronic toluene poisoning. Electrophysiological changes of neuromuscular function in the rats exposed to toluene, *Ind. Med.*, 13, 399, 1971.
194. **Zeyer, H. G.**, Vergiftungen durch Benzol und seine Homologen, in *Handbuch der gesamten Arbeitsmedizin*, Baaders, E. W., Lehmann, G., and Symanski, H., Eds., Schwarzenberg, Berlin, 1970, 422.
195. **Penati, F. and Vigliani, E.**, Sul problema della mielopatie aplastiche, Pseudo-aplastichemiche de benzola, *Russ. Med. Ind.*, 9, 345, 1936.
196. **Rohr, K.**, *Das menschliche Knochenmark*, 2nd ed., Georg Thieme Verlag, Stuttgart, 1949, 250.
197. **Steinberg, B.**, Bone marrow regeneration in experimental benzene intoxication, *Blood*, 6, 550, 1949.

198. **Rondanelli, E. G., Gerua, G., and Magiliulo, E.,** Pathology of erythroblastic mitosis in occupational benzenic erythropathy and erythremia, in *Bibliothica Haematologica, No. 35,* Rondanelli, E. G., Ed., S. Karger, Basel, 1970.

199. **Moeschlin, S. and Speck, B.,** Experimental studies on the mechanism of action on benzene on the bone marrow (radiographic studies using ³H-thymidine), *Acta Haematol.,* 38, 104, 1967.

200. **Heimpel, H., Frickhafen, N., and Heit, W.,** Aplastic anemia: new trends in pathogenesis and treatment, in Main Lectures 7th Meet. Int. Soc. Hematology, European and African Division, Rozman, C., Reichs, A., Eds., Plenery Sessions of the Meeting of the Barcelona, Spain, September 4—9, 1983, 69.

201. **Kansu, E. and Erslev, A. J.,** Aplastic enmia with hot pockets, *Scand. J. Haematol.,* 17, 326, 1976.

202. **Custer, R. H.,** *An Atlas of the Blood and Bone Marrow,* W. B. Saunders, Philadelphia, 1949, 179.

203. **Vilter, R. W., Jerrald, T., Will, J. J., Mueller, J. F., Friedman, B. I. and Hawkins, V. R.,** Refractory anemia with hyperplastic bone marrow, *Blood,* 15, 1, 1960.

204. **Erbengi, T. and Aksoy, M.,** Electron microscopic studies of the bone marrow in four patients with chronic benzene poisoning, Abstracts Lectures and Symposia, ISH, European and African Div., 4th Meeting, Istanbul, September 5—9, 1977, 232.

205. **Engelhardt, E. W.,** *Arch. Gewerbepathol. Gewerbehyg.,* 1, 499, 1931; as cited in **Savilahti, M.,** *Arch. Gewerbepathol. Gewerbehyg.,* 15, 147, 1956.

206. **Nilsby, I.,** Kronisk benzol-forgiftung inon skoindustien, *Nord. Med.,* 42, 1225, 1949; as cited in **Savilahti, M.,** *Arch. Gewerbepathol. Gewerbehyg.,* 15, 147, 1956.

207. **Forni, A. and Vigliani, E. C.,** Chemical leukemogenesis in man, *Ser. Haematol.,* 7, 211, 1974.

208. **Vigliani, E. C. and Forni, A.,** Benzene and leukemia, *Environ. Res.,* 11, 122, 1976.

209. **Erdogan, G. and Aksoy, M.,** Cytogenetic studies in thirteen patients with pancytopenia and leukaemia associated with long-term exposure to benzene, *New Istanbul Contrib. Clin. Sci.,* 10, 230, 1973.

210. **Hernberg, S., Savilahti, M., Ahlman, K., and Asp, S.,** Prognostic aspects of benzene poisoning, *Br. J. Ind. Med.,* 23, 209, 1966.

211. **Scott, J. L., Cartwright, G. C., and Wintrobe, M. M.,** Acquired aplastic anemia. An analysis of thirty-nine cases and review of pertinent literature, *Medicine (Baltimore),* 38, 119, 1959.

212. **Vincent, P. C. and De Gruchy, G. C.,** Complications and treatment of acquired aplastic anemia, *Br. J. Haematol.,* 13, 977, 1966.

213. **Steinberg, B.,** Bone marrow regeneration in experimental benzene intoxication, *Blood,* 6, 550, 1949.

214. **Kissling, M. and Speck, B.,** Further studies on experimental benzene induced aplastic anemia, *Blut,* 25, 97, 1972.

215. **Freedman, M. L.,** The molecular site of benzene toxicity, in *Benzene Toxicity. A Critical Evaluation,* Laskin, S. and Goldstein, B. D., Eds., McGraw-Hill, New York, 1978; *J. Toxicol. Environ. Health,* Suppl. 2, 1977, 37.

216. **Speck, B. and Kissling, M. A.,** A comparative toxicologic evaluation of benzene, toluene and xylene, *13th Int. Congr. Hematology,* Abstract Volume, Lehmann Verlag, Munich, 1970, 134.

216a. **Dexter, T. M., Heyworth, C. M., and Whetton, A. D.,** The role of haemopoietic cell growth factor (interleukin3) in the development of haemopoietic cells, in *Growth Factors in Biology and Medicine,* Pitman, London, 1985, 129.

217. **Uyeki, E., Ashkar, A., Shoeman, A., and Bisel, T.,** Acute toxicity of benzene inhalation to hemopoietic precursor cells, *Toxicol. Appl. Pharmacol.,* 40, 49, 1977.

218. **Gill, D. F., Jenkins, V. K., Kempen, R. R., and Ellis, S.,** The importance of pluripotential stem cells in benzene toxicity, *Toxicology,* 16, 163, 1980.

219. **Horigaya, K., Miller, M. E., Cronkite, E. P., and Drew, R. T.,** The detection of in vivo hematotoxicity of benzene by in vitro liquide bone marrow cultures, *Toxicol. Appl. Pharmacol.,* 60, 346, 1981.

220. **Tunek, A., Olofsson, T., and Berlin, M.,** Toxic effects of benzene and benzene metabolites on granulopoietic stem cells and bone marrow cellularity in mice, *Toxicol. Appl. Pharmacol.,* 59, 149, 1981.

221. **Baarson, K. A., Snyder, C. A., and Albert, R. E.,** Repeated exposure of C57B1 mice to inhaled benzene at 10 ppm markedly depressed erythropoietic colony formation, *Toxicol. Lett.,* 20, 33, 1984.

222. **Frash, V. N., Yushkov, B. G., Kranlov, A. V., and Skuratov, V. L.,** Mechanism of action of benzene on hematopoiesis (investigation of hematopoietic stem cells), *Bull. Exp. Biol. Med.,* 81, 985, 1976.

223. **Haak, H. L.,** Experimental drug-induced aplastic anemia, *Clin. Haematol.,* 9, 621, 1980.

224. **Kalf, G., Post, G., and Snyder, R.,** Inhibition of RNA synthesis kacrophages by benzene, *Fed. Proc. Fed. Am. Soc. Exp. Biol.,* 44, 1704, 1985.

224a. **Post, G., Snyder, R., and Kalf, G. F.,** Metabolism of benzene and phenol in macrophages in vitro and the inhibition of RNA synthesis by benzene metabolites, *Cell Biol. Toxicol.,* 2, 231, 1986.

225. **Gall, E. A.,** Benzene poisoning with bizarre extramedullary hematopoiesis, *Arch. Pathol.,* 25, 315, 1938.

226. **Rawson, R., Parker, F., and Jackson, H.,** Industrial solvents as possible etiologic agents in myeloid metaplasia, *Science,* 93, 541, 1941.

227. **Port, C. D. et al.,** Myeloid metaplasia as a result of chronic benzene intoxication, *N.Y. State J. Med.,* 65, 2260, 1965.

228. **Aksoy, M.**, Problems with benzene in Turkey, *Reg. Toxicol. Pharmacol.*, 1, 147, 1981.
229. **Biscaldi, G. P., et al.**, Acute panmyelopathies due to benzene. Description of a case with favourable outcome, *Med. Lav.*, 64, 363, 1973; English abstract, *Ind. Hyg. Dig.*, 1109, 1974.
230. **Bowers, V. H.**, Reaction of human blood forming tissues to chronic benzene exposure, *Br. J. Ind. Med.*, 4, 87, 1947.
231. **Dacie, J. V.**, *The Haemolytic Anaemias — Congenital and Acquired, Part IV. Drug-induced Haemolytic Anaemia*, 2nd ed., Churchill Livingstone, London, 1967, 1128.
232. **Croizat, P., Guichard, A., Revol, L., Creyssel, R., and Meunier,** Étude anatomo-clinique et chimique, d' un case de mortelle de Marchiafava-Micheli, *Sang*, 19, 218, 1948.
233. **Marchal, G., Leroux, M. E., and Duhamel, G.**, Deficience en un profacteur plasmatique et serique de la thromboplastine (profactor C) au cours d'un syndrome de Marchiafava-Mitcheli, *Sang*, 30, 181, 1959.
234. **Akman, N., Domanic, N., and Müftüoğlu, A. V.**, Paroxysmal nocturnal hemoglobinuria: report of a case with features of intravascular consumption coagulapathy, *Hacettepe Bull. Med. Surg.*, 4, 159, 1971.

Chapter 6

BENZENE CARCINOGENECITY

Muzaffer Aksoy

TABLE OF CONTENTS

I. BENZENE AND LEUKEMIA

As it was noted in the foregoing chapters, although the use of benzene started around the second half of the 19th century, the chronic toxic effect of this chemical agent was seriously considered since 1897, following the description of nine cases of aplastic anemia, four of them fatal, by Santesson[1] in Sweden. Curiously, in the same year, Le Noire and Claude[2] in Paris reported a case of possibly acute leukemia due to chronic exposure to benzene. The patient was a 27-year-old dyer and exposed occupationally to benzene. He showed an acute anemia associated with the findings of hemorrhagic diathesis. Blood counts were not made in this case and the bone marrow was not studied at the autopsy. The occurrence of high amounts of leukocytes in the peripheral blood was taken as evidence for leukemia. The study of the liver also revealed that the trabecular capillaries were full of leukocytes. In the discussion part, Le Noire and Claude mentioned the study of Santesson on benzene-mediated aplastic anemia.* The second and more carefully studied example of this hematologic malignancy, acute lymphoblastic leukemia, was reported by Delore and Borgomano[3] in 1928. Since then, numerous single case reports and small and large case groups associated with benzene exposure have appeared in the literature. Among large groups, the following deserve mentioning: Vigliani and Forni, and Forni and Vigliani in Italy; Goguel et al. in France; Tareef et al. in the U.S.S.R.; Aksoy et al. in Turkey; and the groups surveyed by Browning.[4-10]

It is not possible to calculate the exact number of leukemic patients with benzene exposure, because single cases have been included in collected groups. This point is emphasized clearly in the DFG booklet on benzene.[11] In addition, it is certain that numerous cases of leukemia with chronic exposure to benzene have either not been reported or escaped detection. The last fact has been observed frequently by the present author in his country, either due to insufficient investigation or to a lack of understanding of the role of benzene in the etiology of malignant diseases.

As can be seen in Chapter 2, until 1977 the benzene experiments on animals, apart from trials of Lignac,[12] have not yielded satisfactory results that would lead to a ready acceptance of the leukemogenic effect of benzene. Unfortunately, the only animal experiments showing the leukemogenicity of benzene were the trials of Lignac. But as frequently pointed out, the absence of concurrent controls makes interpretation of the findings of Lignac difficult. These unsatisfactory and conflicting results concerning benzene leukemia prompted the present author and associates since 1970 to perform epidemiological studies in workers chronically exposed to benzene.

A. Incidence of Leukemia Among Shoe Workers Chronically Exposed to Benzene in Istanbul During the Period of 1967 to 1975

As we explained in Chapter 5, prior to 1955 to 1960, benzene was rarely used as a solvent or adhesive by the shoe workers in Turkey.[13] An overwhelming majority of shoes in Turkey, including Istanbul, were produced in small workshops, where usually less than ten persons were employed. Probably it is in some extent similar to the former cottage shoe industries in some of the Italian towns such as those around Milano and Pavia. At that time, the shoe makers were preparing the glue adhesives by processing rubber in petroleum. As they experienced that benzene-containing glue adhesives were extremely practical and cheaper, they replaced their customary adhesives with this new product. At that time the benzene-containing adhesives were produced in very simple and small shops or plants. Consequently, since 1961, numerous cases of aplastic anemia have been noted.[14]

---

* The paper of Le Noire and Claude was ignored by several investigators dealing with benzene leukemia, possibly due to an insufficient study of the case.

Again, up to 1972 the present author and associates collected 44 cases of benzene-mediated aplastic anemia, among whom were a number of workers from small shoe manufacturing workshops.[14,15] The benzene content of 98 samples of adhesives available during 1970 to 1972 in Turkey was determined by gas chromatography in the laboratory of the Turkish Department of Labor (IŞGÜM), Angora. It ranged between 9 and 88% (an average of 50%).[16] Furthermore, from 1967 to 1975 40 leukemic patients with benzene exposure were admitted to the hematology departments of Medical Schools in Istanbul, 34 of the cases occurring among 28,500 shoe workers.[8,9,17] The crude incidence of leukemia among this group was 13.59/100,000 which is significantly higher ($p > 0.01$) than the incidence of 6/100,000 in the general population. A peak of incidence of leukemia (21.7/100,000) occurred between 1971 and 1973.[8,9,17] The possible role of cholera vaccination as an additional leukemogenic factor is discussed later. Nevertheless, our calculation of the incidence of leukemia was probably underestimated since we were not able to include the leukemic shoe workers who may have been admitted to other hospitals in Istanbul.

On the other hand, these studies of the present author and associates have been criticized at certain points.[18,19] For example, according to Goldstein,[18] "the approach used by Italian and Turkish investigators has the advantage of a reasonably high degree of accuracy in the diagnosis, which is the numerator in the risk equation. A definitive disadvantage is that the number of workers at risk, i.e., the denominator, is poorly quantified. This led to much criticism of the Turkish studies, particularly as there are no good records of the number at risk in what is essentially a cottage industry. However, the Aksoy et al. study probably grossly underestimated the relative risk in that they failed to age adjust their findings and the observed cases developed their disease far younger than expected."[18]

Furthermore, according to White et al.,[19] "the representativeness of these few measurements to the entire shoe-working industry, on which the incidence rates are based, is impossible to judge. Therefore, risk values and exposure measurements based on Aksoy's were considered too speculative to be included in a quantitative risk assessment." The results of our epidemiological studies on benzene leukemia in Istanbul may not be considered as "speculative" because of the following facts:

1.   The concentration of benzene in 35 working environments where most of the leukemic individuals worked was measured by a multigas detector and found between 150 to 650 ppm.[20]
2.   The benzene content of the 98 adhesives and thinners available between 1970 and 1972 in Istanbul was determined by gas-chromatography and ranged between 9 and 88% (mean: 50%).[16]
3.   A hematological study among 217 normal-appearing workers in Istanbul was performed. All of the leukemic shoe workers belonged to these workplaces where the study was conducted. As explained in Chapter 5, 24.3% of the workers studied showed hematological abnormalities characteristic for chronic benzene toxicity.[21]
4.   34 patients with aplastic anemia due to chronic exposure to benzene belonged also to the same workplaces as the leukemic shoe workers diagnosed.[14,21,22] The trade names of the solvents used in the workshops of the leukemic shoe workers have been obtained by careful inquiry. These were checked by a list, which showed the benzene content in these solvents, by leukemic workers in Istanbul.[16]

The number of the leukemic workers decreased in 1974 to 1975 to the level of 1969 to 1970 and none were reported in the subsequent 3 years.[9] However, in the period of 1979 to 1985 we have observed 14 new patients with acute leukemia, one with chronic myeloid leukemia, one with hairy cell leukemia, and two with chronic lymphoid leukemia associated with chronic benzene toxicity. Only 2 of these 18 leukemic individuals were shoe workers living in Istanbul. The decline in the annual occurrence of leukemia among shoe workers in Istanbul may be attributed to the gradual prohibition and discontinuation of the use of benzene in Istanbul from 1969 and onwards.[9] The substitutes for benzene in the shoe manufacturing shops were analyzed by gas chromatography. The results showed that several of them contained none or low percentages of benzene such as between 0 to 7.65%.[23] Similarly, the concentration of benzene measured by a Dräger® Multigas detector in some workplaces where previously high concentration of benzene had been found, was mostly now below 10 ppm. On the other hand, the reappearance of leukemia in 1979 and the subsequent 7 years may be attributed either to the variation in the interval between occurrence of leukemia and exposure, or the continuing use of solutions of benzene as a "thinner" and also in solvents.[23-26] We have the impression that both of these possibilities were responsible in the development of leukemia in 18 individuals following 1979 and onwards. On the other hand, the decline in the annual number of leukemic shoe workers after substitution of other compounds for benzene in Istanbul supports the assumption of the leukemogenic effect of benzene on the shoe workers studied.[9,24-26]

Recently, we have observed two cases of acute leukemia in a 6-year period in a modern tire cord factory near Izmit, close to Istanbul.[23,26] In this plant approximately 550 workers were employed. The working conditions were usually good, and the workplaces were large and properly ventilated. The concentration of benzene measured by gas chromatography in one place in the plant was nearly 110 ppm. One of the solvents used in the auxiliary repair shop had a benzene content of nearly 5%.[23,26] Thus, there was an incidence of leukemia (60.6/100,000) in the tire cord plant studied.[23]

B. Incidence of Leukemia Among Other Benzene-Exposed Workers

Since 1938, Vigliani and colleagues[27-30] have published most convincing case reports of leukemia associated with benzene exposure in Milan and Pavia. Occupational exposures were identified in retrogravure plants and in cottage shoe factories. Benzene concentrations near retrogravure machines were 200 to 400 ppm, and benzene concentrations in the air near workers handling glue in shoe factories were 25 to 600 ppm. Estimated latency time from the onset of exposure to clinical diagnosis of leukemia varied between 3 and 24 years. Vigliani and associates[29-31] calculated that for workers heavily exposed to benzene in the provinces of Pavia and Milan, the risk of acute leukemia was at least 20 times higher than that for the general population. According to the working group of IARC, "the study of Vigliani suggests that a relative risk of 20:1 for heavily exposed workers (200-500 ppm), with a median tumour latency of nine years. This appears to be quantitatively compatible with Infante-Rinsky values (4 times higher risk and 2-5 times higher exposure rate."[31,32]

Ishimaru et al.[33] performed a case control study in leukemia in Hiroshima and Nagasaki in order to determine if occupational factors contributed to the incidence of leukemia in these Japanese cities during the period of 1945 to 1967. Of the 492 leukemia cases identified, information could be obtained only for 413 matched case-control pairs. One control per case was chosen from the atomic bomb registry sampling frame and was matched for city, date of birth (± 30 months), distance from the atomic explosion, and residence in either Hiroshima or Nagasaki at the time of disease onset. The risk of leukemia was found to be significantly higher (about $\times 2.5$) among those with a history of occupations in which there was frequent contact with benzene or X-rays, especially among those who had been engaged in these occupations for at least 5 years. Examination in more detail revealed that the risk was higher in those occupations in which various volatile organic solvents were used. On the other hand, as rightly pointed out, the data provides no insight as to whether the effects of these agents are additive, synergetic, or otherwise coleukemogenic.[34]

In 1972, Vidanna and Bross[35] reported the results of a tri-state survey among 1345 men with leukemia and 1237 controls collected in New York (except for New York City),

Baltimore, and Minnesota during the period of 1959 to 1962. According to Vidanna and Bross, preliminary analysis suggested a slightly higher risk of leukemia among workers in the construction industry. Further study showed that the painters group was primarily responsible for this increased risk. Unfortunately Vidanna and Bross were not able to incriminate which chemical or agent was responsible for the increased rate of incidence among painters studied.

A mortality study in a cohort of 13,570 male workers in Akron was described by Monson and Nakano[36] and later Monson and Fine.[37] The potential period of follow-up was from the beginning of 1940 to 1976. The occurrence of cancer, including leukemia, was measured by death certificates. The incidence of lymphosarcoma, reticulum cell sarcoma, and leukemia was seen to be excessive in men who worked at least 5 years in tire making. Excess of leukemia, mostly lymphocytic, was found in many departments where solvents were used. The rate of leukemia in each of these departments was at least 2.5 times the rate in the residual workers. According to the investigators, the most common solvents used in this plant were petroleum solvents, principally high-test gasoline and varnish rubber naptha. All refined petroleum solvents are complex mixtures of hydrocarbons and contain a low percentage of benzene. For the following reason Monson and Fine hesitated to attribute the excess of leukemia solely to benzene; the prominent leukemia was lymphocytic. One of the areas with excess of leukemia deaths also has an excess of lymphosarcoma. McMichael et al.[38-41] reported the results of case-control studies of a major tire manufacturing plant. A cohort of 6678 male workers in this rubber industry was followed for years and data on 1783 deaths were obtained. These studies on the rubber industry workers indicated an association between leukemia and jobs entailing exposure to solvents. It was concluded that the risk of death from lymphoid leukemia appears to be approximately twofold for workers in the medium or low solvent exposure jobs, compared with unexposed. Interestingly, according to McMichael et al.,[38-41] it was chronic lymphoid rather than myeloid leukemia that appeared to be associated with solvent exposure work in this rubber industry. Modern solvents used in the rubber industry supposedly include xylene, toluene, trichlorethylene, and various aliphatic hydrocarbons. Despite this, it was suggested by the investigators that the leukemogenic agent in these solvents might still be benzene.

In a study on the same cohort previously examined by McMichael et al.,[38-41] Andelkovic et al.[42] distinguished between active worker mortality experience and retired worker mortality experience, particularly among the workers who retired before the age of 65. Results were essentially the same as reported by McMichael et al. with findings of excess mortality for both active and retired workers from malignancies of the lymphoid hematopoietic tissues. Thorpe[43] carried out an epidemiological survey of leukemia on 38,000 refinery workers employed in petroleum refining and petrochemical plants in eight European countries. Most of these workers had been exposed to low levels of benzene, much less than the current ceiling value of 25 ppm, over a period of more than 5 years; 18 deaths from leukemia were found among these workers. Thorpe[43] concluded that the occurrence of leukemia in this group of petroleum workers was the same as that of the general population in the countries concerned.

However, this study has been criticized for its relaxed case-finding techniques and its methods of analysis. Many companies had no mechanism which assured that the death of a worker or the cause of death would be reported to the company. Therefore, some deaths due to leukemia were not included in this study.[44,45] Similarly, we are of the opinion that Thorpe was not entirely objective in evaluating the problems concerning benzene leukemia. As an example, he pointed out that in none of the 217 workers with chronic exposure to benzene belonging to the present author and associates series was leukemia observed.[17,21,25] Despite this conclusion, in one worker of this series, 4 years following our study, acute erythroleukemia developed.[17,21,25]

In a small case-control study, Brandt et al.[46] reported on 50 cases of nonlymphocytic leukemia admitted to the university clinic of Lund, Sweden. Of these, 36% had been occasionally exposed to petroleum products or to their combustation products. In 1977, Infante et al.[32] reported the results of an epidemiological study on cohorts of rubber workers exposed to benzene in the course of the manufacturing of pliofilm at two Ohio plants during the period of 1940 to 1949, and followed-up for vital status up to 1975. In comparison with two control populations, a significant excess of leukemia was found in the benzene-exposed workers; a fivefold excess death from myeloid and monocytic leukemia combined. The environment of the workers in the study population was not contaminated with solvents other than benzene and existing records indicate that the benzene levels were generally below the limits recommended at the time they were measured, in most instances ranging from 0 to 10 ppm. In a more recent report of this study, Rinsky et al.[47] reported that most of the 748 workers had been exposed to benzene for a relatively short period of time; 58% of this group had been employed for less than 1 year. According to Rinsky et al., when the data were analyzed by length of employment, a significant excess in leukemia was observed among workers employed for 5 or more years, but not among workers employed for less than 5 years. Among the latter group, two workers had died from leukemia, compared to 1.02 expected, an excess which was not statistically significant. On the other hand, among workers employed more than 5 years, five died from leukemia, compared to 0.23 expected, yielding an SMR of 2100. Infante et al.'s study was criticized by Tabershaw[48] in that the exposure levels did not seem to reflect the facts. According to Tabershaw, measurements taken later show that actual exposure was higher, ranging from 120 to 1000 ppm.

Ott et al.[49] reported the mortality experience of 594 workers occupationally exposed to benzene in the chemical industry, using a retrospective cohort analysis of the period of 1940 to 1973. All workers were employed at the plant between 1940 and 1970. Three of these workers died from leukemia; they were classified as one, myelocytic and two, acute myelocytic leukemia. The increase in leukemia death rate was of borderline statistical significance. The time-weighted average benzene exposure of these three individuals was below 10 ppm. Furthermore, De Couflé et al.[50] performed a historical cohort mortality study comprising 259 employees of a chemical plant where benzene had been used in large quantities. The study group included all persons between January 1, 1947 and December 31, 1960. The cohort was followed through December 31, 1977. The only unusual finding was four deaths from lymphoreticular cancers when 1.1 would have been expected on the basis of national mortality rates. Three of those deaths were due to leukemia and one was by multiple myeloma. According to the investigators, these findings are consistent with the previous reports of leukemia following occupational exposure to benzene, and raise the possibility that multiple myeloma could be also linked to benzene.

Girard and Revol[51] conducted a study in hospitals of Lyon by questioning the patients, analyzing solvents used, and making inquiries at the work site. In this retrospective study among 401 patients with hematological disorders, such as acute lymphoblastic leukemia, chronic lymphocytic leukemia, and aplastic anemia, 46 individuals were found with a history of exposure to benzene and toluene. As compared with the values reported in controls, this frequency was statistically significant. Contrary to this, the leukemogenic effect of toluene is generally not accepted. Vigliani and Forni[4] reported that since the replacement of benzene with toluene as a solvent in the retrogravure industry in Italy, they have seen no cases of aplastic anemia, nor leukemia, due to toluene exposure. Furthermore, toluene-exposed workers did not show the chromosome alterations frequently seen in workers exposed to benzene.[5]

Considering mortality excess from lymphocytic leukemia due to solvent exposures, Arp et al.[53] performed a detailed examination of the solvent history experiences of the 15 cases with lymphocytic leukemia previously reported by Michael et al.[38-41] and their matched industrial controls. The results of a detailed examination of the solvent exposure histories of 15 cases of lymphocytic leukemia and 30 matched industry controls showed that cases of lymphocytic leukemia were 4.5 times as likely as controls to have had direct exposures to benzene and other solvents. Analysis of manufacturing production source of solvents revealed that cases spent greater proportions of their work experience in jobs with potential exposure to coal-based benzene and xylene. No differences were seen for petroleum-derived solvents. Cases with lymphoid leukemia were 6.67 times more likely than controls to have been exposed to coal-based solvents; the corresponding relative risk for petroleum-based solvents was 1.5. Despite this, the investigators were hesitant to accept the causal interferences regarding lymphocytic leukemia and coal-based solvents, or their contaminants, which cannot be drawn conclusively from the frequency of their study.

Considering excessive leukemia mortality among British and U.S. rubber industries, recently Chekoway et al.[54] attempted to identify causative factors in 11 workers whose underlying cause of death was lymphocytic leukemia. They were identified from the same cohort reported on by McMichael et al.[38-41] These cases were also included in the study of Arp et al.[53] A total of 1350 controls were included in the study. The case-control analysis of lymphocytic leukemia studied was extended to consider risks associated with 24 solvents used in this cohort. These were benzene, carbon disulfide, carbon tetrachloride, gasoline, heptane, hexane, isopropanol, methanol, perchlorethylene, phenol, special napthas, toluene, trichlorethane, trichlorethylene, xylene, etc. Solvent exposure charts for process areas were reconstructed from historical plant records. Chekoway et al.[54] concluded that the associations with lymphocytic leukemia risk observed for the solvents mentioned above, most notably carbon tetrachloride and carbon disulfide were stronger than those detected for benzene.

C. Types of Leukemia

Since the first reports of leukemia due to chronic exposure to benzene, numerous single case reports and small and large case groups with benzene exposure have been appearing in the literature.[4-10] On the other hand, it is apparent from the literature that there is a significant difference concerning the distribution of the types of leukemia in the groups mentioned above.[4-10] In one group, acute types of leukemia predominate whereas in the other series, chronic types of leukemia take the most important place.

Recently, Yin et al. published the results of a retrospective cohort study in 233 benzene and 83 control factories in 12 cities of China.[54a] The benzene cohort and the control cohort consisted of 28,460 benzene exposed workers (178,556 person-years from 1972 to 1981) and 28,257 control workers (199,201 person-years) in the latter. The leukemia mortality rate was 14/100,000 person-years in the benzene cohort and 2/100,000 person-years in the control cohort. The average latency of benzene leukemia was 11.4 years. Most of the workers (76.8%) were of the acute type. The mortality due to benzene leukemia was highest in organic synthesis plants, followed by painting and rubber synthesis industries. The concentration of benzene to which patients with leukemia were exposed ranged from 10 to 1000 mg/m^3 (mostly from 50 to 500 mg/m^3). Of the 25 cases of leukemia, 7 had a history of chronic benzene toxicity prior to the occurrence of leukemia.

D. Distribution of the Types of Leukemia

To illustrate the above-mentioned important differences, we have summarized the distribution of the types of leukemia with chronic benzene toxicity in some groups, including ours and those of 50 nonexposed leukemic individuals, in Table 1. In our series, 52 out of 58 leukemic individuals (89.6%) had some forms of acute leukemia. According to this study, acute myeloblastic leukemia (AML) seems to be the most frequently encountered following chronic benzene toxicity (39.7%). This finding is in accordance with those of Vigliani and Forni[4,5,30] and those of Infante et al.[32] Acute myeloblastic leukemia was followed in frequency by acute erythroleukemia (17.3%) and preleukemia (12%). On the other hand, acute mye-

Table 1
THE DISTRIBUTION OF LEUKEMIA TYPES IN SOME SERIES OF PATIENTS WITH CHRONIC BENZENE TOXICITY

	Vigliani et al.[4]		Goguel et al.[6]		Infante et al.[32]		Aksoy et al.[8,9] Exposed		Aksoy et al.[8,9] Nonexposed	
Types of leukemia	No.	%	No.	%	No.	%	No.	%	No.	%
AML	19	82.6	11	25.0	4	57	23	39.7	8	16
ALL			2	4.5			5	8.6	13	26
Preleukemia			5	11.4			7	12	1	2
A. erythroleukemia	4	17.4	2	4.5			10	17.3	2	4
A. myelomonocytic l.			1	2.3	2	28.5	5	8.6	3	6
A. undifferentiated l.			2	4.5			1	1.7		
A. premyelocytic l.							1	1.7		
Hairy cell l.							1	1.7		
CML			13	29.6	1	14.5	3	5.2	10	20
CLL			8	18.2			2	3.5	13	26

lomonocytic leukemia and acute lymphoblastic leukemia (ALL) were comparatively rare (8.6%). In the series of Infante et al.,[32] acute myeloblastic leukemia was the most frequent type of leukemia, followed by acute myelomonocytic leukemia. In contrast to the results of these series, there are several reports describing the frequent occurrence of chronic myeloid (CML) or even chronic lymphoid leukemia (CLL) following chronic benzene toxicity. Browning[10] was able to collect 61 cases of leukemia associated with chronic benzene exposure from the literature; 16 (26.2%) had chronic myeloid leukemia and 6 (9.8%) showed the findings of chronic lymphoid leukemia. Similarly, Tareef et al.[7] described 16 leukemic workers with exposure to benzene; six (37.5%) had acute leukemia, five (31.25%) showed the findings of chronic myeloid leukemia, and three (18.7%) had chronic lymphoid leukemia. Furthermore, if we compare the types of leukemia occurring in our 58 patients belonging to our series and 50 nonexposed persons, we will see striking differences in the distribution of the types of leukemia. For example the latter group showed chronic lymphoid leukemia in 26% of the individuals while only two (3.5%) of the exposed group showed such findings. More important is that 46% of the nonexposed group had chronic leukemia in contrast with the 10.4% in the exposed group.

This result is also in contrast with that observed among the heavily irradiated survivors in Hiroshima and Nagasaki.[55] The occurrence of chronic myeloid leukemia was frequent among these groups of survivors. Furthermore, a small but significant difference was found concerning the incidence of chronic myeloid leukemia in atomic bomb survivors in Hiroshima and Nagasaki, being 33% and 18%, respectively.[56] This difference in the incidence rate of chronic myeloid leukemia was found to be related to the dose and nature of the radiation.[56] The almost complete absence of chronic lymphoid leukemia among these groups of survivors of the atomic bomb explosion was explained by the infrequent occurrence of chronic lymphoid leukemia in Japan.[57] As can be seen from Table 1, the distribution of the types of leukemia in Turkey is quite similar to those observed in Western countries.[8,58] Similarly, in a survey concerning a 10-year period, among 3715 hematologic patients at the hematology section, Department of Internal Medicine, Istanbul Medical School, 695 patients had leukemia; 396 (57%) of the leukemic patients had acute leukemia, and 299 (43%) had chronic leukemia.[58] Therefore, the very low percentage of chronic lymphoid leukemia and the rarity of chronic myeloid leukemia in our series is significant. Furthermore, the distribution of the types of leukemia obtained in 58 leukemic patients with long-term exposure to benzene is similar to that observed by Court-Brown and Doll[59,60] in patients receiving partial body irradiation for the management of ankylosing spondylitis and the frequency of acute types of leukemia encountered in patients receiving radioiodine for the treatment of hyperthyroidism.[61,62]

In the Court-Brown and Doll report,[59,60] 38 out of 50 cases had acute leukemia (76%) and only eight patients (16%) with chronic forms, myeloid and lymphoid, were found; the remaining four had unidentified types of leukemia. There are several reported cases of leukemia following the radioiodine treatment. With the exception of two, all had acute leukemia.[61,62]

1. Acute Myeloblastic Leukemia

Acute myeloid leukemia is the most frequent of benzene-mediated leukemia, both in the larger group of patients and single case reports due to chronic benzene toxicity. Most of the cases with acute myeloid leukemia progress as the aleukemic type. Leukocytes can be as low as $0.5/10^9/\ell$, as was observed in one of our patients.[63] The patient, a shoe worker with 6 years of benzene exposure, had acquired acute myeloblastic leukemia 3 years after recovery from aplastic anemia. In 11 out of 23 patients with acute myeloid leukemia, the leukocyte count ranged between 0.5 to $3.5/10^9/\ell$. In our series, acute myeloid leukemia was found in 23 patients (39.7%) and it was encountered in 19 out of 23 cases (82.7%) reported by Vigliani and Forni.[4] Contrary to these, acute myeloid leukemia was diagnosed in 5 out of 16 patients with chronic benzene toxicity (31.25%) reported by Tareef et al.,[7] and in 11 out of 44 cases in the Goguel et al. series[6] (25%).

Recently, acute myeloid leukemia in man has been considered as a disease primarily involving a block in the normal differentiation of myeloid precursors.[64] According to Ruscetti et al.,[64] the etiology of this block remains obscure, but ionizing radiation, chloramphenicol, benzene exposure, and type C-retroviruses have all been implicated.

2. Acute Lymphoblastic Leukemia

Following the first case of Delore and Borgomano,[3] several cases of acute lymphoblastic leukemia have been reported. In our series, it was found in five cases (8.6%), and like acute myeloid leukemia, significant proportions of these were aleukemic. It was found in two (4.5%) in the Goguel et al. series.[6] It is not always possible to differentiate cell types in benzene-mediated acute leukemia unless cytochromic studies are performed. Unfortunately, in most of the reported cases, including our series, this method is not performed. This is the reason for the difficulty encountered in positively identifying whether the type of leukemia is myeloid or lymphoid. The case of Duvoir et al.[65] is an example. An employee died of severe anemia after having worked with benzene for 20 months. Findings that would indicate acute leukemia were seen in autopsy. Evidences of hemohistioblastosis were found in the bone marrow, spleen, and lymphoid nodes. It was impossible to differentiate cell types. According to the investigators, no evidence of leukemia was seen in peripheral blood. Similarly, Goguel et al.[6] have pointed out the difficulty in differentiating blasts in benzene-mediated leukemias. The possible case of acute lymphoblastic leukemia reported by Hernberg et al.[66] developed after 9 years following pancytopenia due to chronic benzene exposure. Furthermore, Browning[10] encountered a case of acute lymphoblastic leukemia in a worker from an artificial leather shop.

3. Acute Erythroleukemia

Acute erythroleukemia has been reported to be one of the most frequent types of leukemia caused by benzene. Acute erythroleukemia was found in 10 out of 58 patients with chronic benzene toxicity (17.3%) in our series (Table 1). According to Forni and Vigliani,[5] up to 1974, at least 20 cases of acute erythroleukemia due to chronic exposure to benzene have been reported. Among their series, four patients (17.4%) showed the findings of acute erythroleukemia, and it was established in 2 out of 44 patients of Goguel et al.[6] as 4.5%.

In view of the rarity of this disease, they considered this low number of cases to be significant. In the nonexposed group of our series only 2 out of 50 patients (4%) had this type of leukemia.[8] In 1950, Di Guglielmo and Iannacone[67] encountered a case of erythremic myelosis in a man who printed colored pictures with a retrogravure machine. In whom, it was reported that the mitotic activity was inhibited at the metaphase, and degenerative changes had occurred in the nucleus only. The investigators stated that acute benzene poisoning caused acute benzene erythremia. In the Tareef et al.[7] series, there were two cases of erythroleukemia (12.5%), one of acute type. In his testimony to OSHA (Occupational Safety and Health Administration) Sakol,[68] an Akron hematologist, described what he considered to be an epidemic of a rare form of leukemia among certain workers. From 1954 to 1963 he had observed nine cases, including at least four and probably nine of erythroleukemia, all possibly exposed to benzene during a pliofilm operation.

4. Benzene Erythroleukemia Manifesting Delta Beta-Thalassemia

Markham et al.[69] reported a 27-year-old male with aplastic anemia due to benzene exposure. The patient developed a high level of fetal hemoglobin, ranging between 12 and 40%, a low content of hemoglobin A_2 (0.85%), a decreased β/α synthesis ratio (0.46 to 0.39), and increased $G\gamma/A\gamma$ synthesis ratio (0.99).[69] This acquired hemoglobinopathy resembling δ beta-thalassemia was recognized at the onset of acute erythroleukemia. According to the investigators, certain features of this abnormal globin synthetic pattern resemble those of the normal fetus and thus appear to provide another example of gene expression by malignant cells resembling that of an earlier stage of the organism's development.

Here, we should emphasize that all these findings resembling δ beta-thalassemia are not specific for chronic benzene toxicity. Also, they were usually encountered in nonexposed cases of acute erythroleukemia.

5. Acute Monocytic and Myelomonocytic Leukemia

The first case of acute myelomonocytic leukemia was reported by Marchand[70] in a worker who was applying benzene-containing paint to metal furniture. In our series, five patients with acute myelomonocytic leukemia (8.6%) were found. In one of them, a 42-year-old shoe worker, acute myelomonocytic leukemia developed 2 years following diagnosis of aplastic anemia.[63] He was exposed to benzene for 10 years. He had pancytopenia and hypocellular bone marrow. At that time he was not transfused. His pancytopenia subsided following steroid therapy and the patient continued to work until a short time before his admission to our clinic, thus being chronically exposed to benzene. Approximately 2 years after occurrence of aplastic anemia, thrombocythemia appeared as a preleukemic stage with such symptoms as tenacious bone and muscle pains, irregular fever, moderate anemia, and some atypical monocytes in the peripheral blood. The results of the bone marrow puncture were not consistent with the diagnosis of leukemia. There were numerous megakaryocytes in the bone marrow. The platelet count ranged between 540 to $1600/10^9/\ell$. A second bone marrow puncture about 3.5 months later, showed changes characteristic of acute myelomonocytic leukemia, namely 9% myeloblasts, 20% atypical monocytes, and few Auer's bodies in the monocytes. There was also marked lysozymuria. In the Infante et al.[32] series, acute myelomonocytic leukemia was found in two patients (28.5%) and in one case in the Goguel et al.[6] series (2.3%).

6. Preleukemia

The high frequency of preleukemia in some series of benzene leukemia is another striking feature of chronic benzene toxicity. The term, preleukemia, is used for any hematologic syndrome that may in time develop into overt leukemia, but which lacks the criteria essential for the diagnosis of overt leukemia when the patient is first seen. Our criteria for diagnosis of preleukemia were similar to those of Wintrobe et al.[71]

1. The presence of refractory anemia, mostly hypochromic and often associated with thrombocytopenia or pancytopenia.
2. Normal ranging or slightly increased percentages of myeloblasts in the bone marrow.
3. The presence of a few blast cells, mostly myeloblasts, in the peripheral blood film.

Preleukemia was found in the Goguel et al.[6] series in five patients (11.4%) and in our series in seven (12%). In contrast to this rate, it was 4% in our nonexposed group.[8] The absence of this type of leukemia in some series of chronic benzene toxicity possibly is due to nonconsideration of this disorder in case evaluation.

7. Acute Promyelocytic Leukemia

In our series, this type of acute leukemia was observed in only one case, 1.7%.[17] In a 36-year-old shoe worker with 2.5 years exposure to benzene, an acute promyelocytic leumia developed. There was 30% of the promyelocytes in the bone marrow. This patient showed decreased level of serum fibrinogen which is a characteristic finding of this type of leukemia.

8. Chronic Myeloid Leukemia

The first such case was reported by Emil-Weil[72] in a female worker and it was atypical, presenting with symptoms of aplastic anemia and aleukemia at the first examination, then terminating with acute myeloblastic leukemia. Although chronic myeloid leukemia was absent or in low frequencies in some series of benzene leukemia, such as those of the Aksoy et al., Vigliani and Forni, and Forni and Vigliani series, it was as high as 29.5% (in 13 patients) and 31.25% (in 5 patients) in the Goguel et al. and the Tareef et al. series, respectively.[4-9] On the other hand, Browning[10] collected 16 patients (26.2%) with the type of chronic myeloid among 61 cases of the literature. Despite this, in our series, chronic myeloid leukemia was found in only three patients (5.2%) whereas in the nonexposed group it was 20% (Table 1).

There are single case reports describing the occurrence of chronic myeloid leukemia due to chronic benzene exposure. For example, Kahler and Merker[73] have described chronic myeloid leukemia in a 57-year-old worker exposed 7 years to benzene. Similarly, Marchal and Duhamel[74] have reported two cases of chronic myeloid leukemia, one of which showed the findings of increased hemolysis. Additionally, Hellriegel et al.[75] reported a case of chronic myeloid leukemia which developed 15 months later following occurrence of aplastic anemia due to chronic exposure to benzene.

Finally, today it is difficult to explain the significant difference between the frequency of chronic myeloid leukemia due to chronic benzene exposure in our and Vigliani and associates series on the one hand, and in the groups reported by the Browning, Goguel et al., and Tareef et al. series on the other.[4-10] This problem will be discussed later.

9. Chronic Lymphoid Leukemia

Two years after the occurrence of aplastic anemia in a 58-year-old worker who had been exposed to benzene, Falconer[76] reported the first case of chronic lymphoid leukemia due to this chemical. According to Falconer, benzene acts like an agent that damages the hematopoietic tissue. It later alters cell production and homeostasis, leading to leukemia. Although this kind of leukemia was absent in the series of Vigliani and Forni and in only two in our series (Table 1)[77,24] consisting of 58 leukemic patients with chronic benzene toxicity, it was present in the Goguel et al. series as 18.2% (in 8 cases), in Browning's collected cases as 13.1% (in 6 patients), in the Tareef et al. series in 18.75% (in 3 cases) and in our nonexposed group it was 26%.[4-10] On the other hand, in epidemiologic studies of McMichael et al. and Andelkovic et al., chronic lymphoid leukemia was predominant.[37-41] Similarly, Girard and Revol[51] have reported a relatively high incidence of chronic lymphoid leukemia often exceeding the number of cases of chronic myeloid leukemia in the retrospective statistical study among 401 patients with hematological disorders.

Considering these data, we have forwarded recently that although in our series and in some similar series there were no cases of chronic lymphoid leukemia, this type of leukemia, even rarely, might be found in chronic benzene toxicity.[4,5,77] Indeed, a short time later, in 1985, we had encountered two possible cases of chronic lymphoid leukemia due to chronic exposure to benzene (Table 1).[24]

Case 56, a 43-year-old Turkish Jew, owner of a wallpaper printing shop in Istanbul had been seen by an internist because of cardiac complaints. The results of EKG were normal. There was no lymphadenopathy and hepatosplenomegaly. Hematologic data: Hb 14.3 g/dℓ, RBC 4.8 × 10^9, PCV 0.45/ℓ, platelet 380 × 10^9/ℓ, WBC 29 × 10^9/ℓ with 3% band forms, 25% polymorphonuclear neutrophils, 2% monocytes, 2% eosinophiles, and 68% lymphocytes. There were numerous Gumprecht shadows in the blood film. The findings of bone marrow were consistent with the diagnosis of chronic lymphoid leukemia. Serum electrophoresis showed albumin 59.4%, alpha-1 globulin 2.2%, alpha-2 globulin 9.9%, beta-globulin 12.7%, and gamma globulin 15.8%.

He never used chloramphenicol and was never exposed to radiation. He was the owner of a wallpaper printing shop for 2 years. For this purpose, he mixes pigmented dyes with a solution of toluene and methyl alcohol keton. He is busy for 2 to 3 hours daily in his plant. According to his statement, he sniffs above mentioned solutions for control purposes. The toluene solution was analyzed by a gas chromatograph (Varian®) and found to contain 2.8% benzene and 95.3% toluene. The other solution of methyl alcohol ketone contained no benzene. Before this job, he was in a shop selling different chemical materials.

The second patient of chronic lymphoid leukemia (case 58) was a 51-year-old owner of a plant where some automobile parts are produced or repaired. Because his room was not far from the active part of the plant, he was intermittently exposed to the solutions used. In this plant some cleaning solutions were also used. They contained no benzene as determined by gas chromatography.

In the period of 1955 to 1965, the patient was the owner of a small plastic plant and similarly, he was intermittently exposed to thinners containing approximately 27.3% of benzene. He had some pains on the right upper quadrant of his abdomen. There was a mild hepatosplenomegaly and no lymphadenopathy. A blood examination revealed leukocytes 23 × 10^9/ℓ with 73% of lymphocytes. The findings of the bone marrow puncture at sternum were consistent with the diagnosis of chronic lymphoid leukemia. If we compare certain epidemiological data of these two cases of chronic lymphoid leukemia and those of 52 patients with acute leukemia in our series, we find out some differences. (1)We find that there is a difference in the content of benzene in the adhesives used. During the period of 1970 to 1972, it was between 9 and 88% (mean: 50%).[16] Despite this, benzene content of the solution which was used by the first patient with chronic lymphoid leukemia (case 56) in our series was as low as 2.8%. (2) The adhesives which were used by nearly all the workers with acute leukemia in our series contained only benzene, but not the other homologues of this chemical such as toluene and xylene. This fact was particularly evident before the period of 1967 to 1976. Contrary to this, the solvent used in the workplace by the first case with chronic lymphoid leukemia in our series contained 95.3% toluene, and the second patient (case 58) was later exposed to different chemicals. (3) The great majority of the individuals with acute leukemia in our series were exposed to high concentrations of benzene, between 150 to 210 ppm, during all working hours. Contrary to this, four out of five patients with chronic leukemia, two with chronic myeloid leukemia and two with chronic lymphoid leukemia, were exposed to benzene intermittently and for a short time during the daily work.

Considering these data, we suggest that the difference concerning the distribution of types of leukemia in the groups from different countries or studies can be explained partly by the exposure levels and also by the presence or absence of other homologues of benzene such as toluene and xylene or other chemicals.[24]

10. Chronic Erythremic Myelosis

Di Guglielmo and Ricci[78] reported a case of this type of leukemia in a 63-year-old female shoe worker who worked with benzene-containing materials. The patient showed severe anemia and 68% of erythroblasts in the peripheral blood smear. The bone marrow findings confirmed the erythremic myelosis. The patient lived 2 years following the diagnosis. In the Tareef et al.[7] series a similar case of chronic erythremic myelosis was observed with 5 years of survival.

11. Hairy Cell Leukemia

There are a few case reports on hairy cell leukemia due to chronic exposure to benzene.[24,79,80] The first case of this type of leukemia was reported by Daily et al.[79] in France. Recently, we have observed one case of hairy cell leukemia possibly due to chronic benzene toxicity.[24] The patient was a 50-year-old manager in a plastic workshop. He was exposed to benzene during the period of 1957 to 1965, and later to polystyrene and polyethylene. In 1984, hairy cell leukemia developed. Because of diabetes mellitus and furonculosis at scrotal and inguinal regions, he was admitted to a hospital in Istanbul. There, a spontaneous spleen rupture occurred and therefore, he was splenectomized. He went to Switzerland and in Kantonspital in Basel he was diagnosed as having hairy cell leukemia. Histochemical test for tartarate-resistant acid phosphatase revealed 16% lymphoid cells. At present, he is in remission by interferon therapy.

E. Preceding Pancytopenic Period

It is often noted that leukemia associated with benzene exposure frequently develops following a period of bone marrow depression. As Browning[10] pointed out, such a transition from aplastic anemia to leukemia is not unknown in the idiopathic type of this hematologic disorder, but it seems that it happens more rarely than encountered in leukemia following benzene exposure. According to Vigliani and Saita[28,29] and Forni and Vigliani,[5] leukemia develops in benzene-induced aplastic anemia or pancytopenia of longstanding; in these cases, during the course of the disease the bone marrow changes from a hypoplastic to an aleukemic pattern. On the other hand, according to Cronkite[81] "the finger of suspicion must be pointed out at any agent which is able to produce an aplasia of the bone marrow assuming it will probably be able to produce leukemia also".

As can be seen from Table 2, a preceding pancytopenic period was present in 13 out of 58 patients (22.4%), five with acute myeloblastic leukemia, four with preleukemia, two with acute erythroleukemia, one with acute myelomonocytic leukemia, and one with acute undifferentiated leukemia.[25] Interestingly, this period was present in 57% of the patients with preleukemia. The interval between the onset of preceding period and that of the manifestation of leukemia varied from 4 months to 6 years. Quite often the clinical and hematological findings of pancytopenia improved considerably, or the condition even disappeared completely. Despite any such improvement, leukemia developed later.

As we explained in Chapter 5, we have performed a follow-up study of 2 to 17 years on 44 pancytopenic patients with chronic benzene toxicity.[15] In six (13.6%) leukemia developed. In addition, in one of these 44 pancytopenic patients, fatal myeloid metaplasia developed after 8 years.[15] The development of leukemia in 6 out of 44 patients is a further evidence of the leukemogenic effect of benzene in man. Sometimes the interval between the onset of preceding pancytopenic period and the occurrence of leukemia may be as long as 14 or 15 years, as observed in the patients of De Gowin[82] and Vigliani and Forni.[4] Furthermore, as

Table 2
PRECEDING PANCYTOPENIC PERIOD PRESENT IN 58 LEUKEMIC INDIVIDUALS ASSOCIATED WITH CHRONIC BENZENE TOXICITY

Type of leukemia	Patients (no.)	%
Acute myeloblastic leukemia	5(23)	21.7
Preleukemia	4(7)	57
Acute erythroleukemia	2(10)	20
Acute myelomonocytic leukemia	1(5)	20
Acute undifferentiated leukemia	1(1)	

Note: The numbers in the parentheses represent the total number of leukemic individuals.

Goldstein[83] pointed out, there is frequently a long delay between the onset of leukemia and the cessation of known benzene exposure. On the other hand, in 6 out of 13 patients with preceding pancytopenic period which later developed into leukemia, corticosteroids were used in the treatment of bone marrow depression. The use of corticosteroids has been suggested to be a contributory factor in the development of leukemia following benzene exposure.[25,63]

F. Duration of Exposure

The correlation between the occurrence of leukemia and the duration of exposure varies widely. It ranged between 4 months and 40 years with a mean of 10.2 in our series. In 25 patients, the duration of exposure was more than 10 years and in 3 of them, it was more than 20 years. The shortest duration of exposure was recorded in one of our patients; following 4 months of exposure, a severe pancytopenia developed and 6 years later, the patient died in preleukemia.[8,15] Another short period, such as 6 months, in a case of acute myeloblastic leukemia was reported by Kinoshita et al.[84] and also developed following the occurrence of preceding pancytopenic period. Similarly, the duration of exposure was as short as 7 months in the patient reported by Erf and Rhodas.[85]

Contrary to these, in a case of acute myeloblastic leukemia reported by Pugni et al.,[86] the duration of exposure was as long as 40 years. The duration of exposure varied between 5 and 17 years with a mean of 10.9 in the Bernard and Braier[87] report on five patients with acute myeloblastic leukemia. In the Tareef et al.[7] series in eight cases of acute myeloblastic leukemia, the duration of exposure varied between 5 and 22 years, between 20 and 22 in two with acute erythroleukemia, between 4 and 27 years in five with chronic myeloid leukemia, and between 10 and 22 in three with chronic lymphoid leukemia. In the Goguel et al.[6] series, the duration of exposure ranged between 5 and 27 years in acute leukemia (mean: 7.3 years), between 1 and 20 years (mean: 6.7 years) in 13 patients with chronic myeloid leukemia, and between 5 and 35 years (mean: 18.1 years) in eight cases of chronic lymphoid leukemia.

All these data clearly show that there is a significant variation in the duration of exposure prior to the onset of benzene-induced leukemia with chronic benzene toxicity. In our series, there was some correlation between the duration of exposure and the types of leukemia (see Table 3). The exposure was shortest in acute lymphoblastic leukemia and the longest in acute erythroleukemia and acute myeloblastic leukemia, but there are several examples of leukemia due to benzene exposure with differing results concerning this relationship.[25] These data show that there is no clear correlation between the duration of exposure and the type of leukemia.

Table 3
DURATION OF EXPOSURE IN DIFFERENT TYPES OF
LEUKEMIA IN 58 PATIENTS ASSOCIATED WITH
LONG-TERM EXPOSURE TO BENZENE

Types	Duration	Mean duration
Acute myeloblastic leukemia	1.5—40 years	11 years, 6 months
Acute lymphoblastic leukemia	6 months—6 years	3 years, 3 months
Acute erythroleukemia	3—25 years	14 years, 1 month
Preleukemia	4 months—15 years	8 years, 2 months
Acute myelomonocytic leukemia	1.5—10 years	4 years
Acute undifferentiated leukemia	20 years	—
Acute promyelocytic leukemia	2.5 years	—
Hairy cell leukemia	8 years	—
Chronic myeloid leukemia	4—12 years	8 years
Chronic lymphoid leukemia	2—10 years	6 years

Table 4
THE RELATIONSHIP BETWEEN FIVE LEUKEMIC
PATIENTS WITH CHRNOIC BENZENE TOXICITY

Case no.	Age (years)	Duration of exposure (years)	Occupation	Type of leukemia	Relationship
1	43	6	Shoe worker	AML	Paternal uncle of case 2
2	24	4	Shoe worker	ALL	Nephew of case 1
3	36	7	Shoe worker	Pre L	Maternal cousin of case 4
4	48	3	Painter	AMYL	Maternal cousin case 3
5[a]	36	15	Shoe worker	AERL	

[a]   The relationship of case 5 is explained in the text.

From Aksoy, M., Erdem, Ş., and Dinçol, G., *Acta Haematol.*, 55, 65, 1976. With permission.

G. Genetic Factors

There are data suggesting that genetic determinants or host factors may play a role in the development of malignancies including leukemia. Several families are reported in which two or more cases of leukemia of the acute type or particularly the chronic lymphoid type occurred in the same or successive generations.[88-91] Furthermore, a striking increase of acute leukemia in twins is well-documented.[92] Leukemia may also be associated with congenital defects such as Down's syndrome (trisomy 21).[93] Additionally, a high incidence of leukemia has been repeatedly reported in several hereditary disorders associated with excessive chromosome breakage such as Fanconi syndrome, Bloom's syndrome, and ataxia telengectasis.[94-96]

As can be seen from Table 4, in our series in five leukemic patients with chronic benzene toxicity, a genetic determinant was established.[8] Two were paternal uncle and nephew and two were cousins. The father of the fifth leukemic patient with acute erythroleukemia, a 65-year-old shoe worker with a long history of exposure, died in the hospital with the diagnosis of myelosclerosis, but reevaluation of the case report showed that this patient possibly had acute leukemia of unidentified type.[8] These five leukemic patients among first and second degree relatives constituted 8.6% of our leukemic series with chronic benzene toxicity. The possibility of a coincidental occurrence among close relatives in our series is unlikely. Therefore, in these five patients, the development of leukemia was possibly due to the presence of genetic determinants or host factors in addition to benzene as an environmental or extraneous factor. In other words, some genetic factors triggered by chronic exposure to benzene could have played a role in these three families. This suggestion is in accordance with the view that leukemia may be due to the combination of various intrinsic and extrinsic factors.[89]

Gallinelli and Traldi[97] reported two sisters with chronic benzene toxicity who developed pancytopenia and bone marrow aplasia during pregnancy. The younger sister was completely cured following delivery. The older sister had begun to improve following a miscarriage, but she became pregnant again and this time acute leukemia developed. These findings are difficult to interpret, but do not directly support a hereditary influence.

H. The Concentration of Benzene in Workplaces

Unfortunately, exact and reliable data showing the correlation between the concentration of benzene in the air at the workplace and the occurrence of leukemia is lacking. It has not been possible to establish a quantitative relationship between the concentration and the duration of exposure to the development of leukemia from the histories of single cases, because they are too small in number and the concentrations stated are unreliable.[11] As an example, a case of acute myelosis had been initially attributed to benzene exposure over 4 years at a concentration exposure of possibly less than 25 ppm.[98] But subsequent studies revealed that this leukemic individual had been subjected to higher concentrations of benzene prior to the quoted period.[99] In contrast to this, in the case reported by Stieglitz et al.,[100] leukemia developed following pancytopenia in a 26-year-old individual 5 years after working with a fluid used to clean machine parts which contained 4 to 6% benzene. The level at the working environment was as low as 20 ppm. Similarly, there was a discussion on the benzene levels recorded by Infante et al. in two rubber factories.[101,102] Tabershaw and Lamm[101] argued that exposures in these workshops were not as low as Infante et al. had stated, and pointed out occasional high excursions occurred in airborne benzene levels, up to several hundred ppm. Contrary to this, Rinsky et al.[47] stated that most of such high excursions occurred in areas entered only infrequently by workers.[103]

I. Some Possible Factors in the Development of Benzene Leukemia

1.   The possible role of corticosteroids: in our series of benzene leukemia, in 6 out of 11 patients with preceding pancytopenic period in whom leukemia developed, corticosteroids were used and the use of corticosteroids has been suggested to be a contributory factor in the development of leukemia following benzene exposure.[15,25,63]
2.   The possible role of reexposure to benzene: reexposure to benzene as a factor in the development of leukemia was noted in four out of nine leukemia cases with a preceding pancytopenic period.[15,25,63] For the present author, the most effective factor in the development of leukemia, following pancytopenia or hematological abnormalities due to chronic benzene toxicity, is the reexposure to this chemical.
3.   The possible role of thalassemia in the development of leukemia in the individuals with chronic benzene toxicity: among our series of benzene-leukemia, a patient with acute myeloblastic leukemia (or erythroleukemia) had a hypochromic microcytic anemia with normal levels of hemoglobins A_2 and F.[17,104,105] Two children of this leukemic patient were found to have a hypochromic microcytic anemia with normal levels of

Table 5
ANNUAL NUMBER
OF LEUKEMIC SHOE
WORKERS IN
ISTANBUL BETWEEN
1967 AND 1978

Years	No. leukemic shoe workers
1967	1
1968	1
1969	3
1970	4
1971	6
1972	5
1973	7
1974	4
1975	3
1976	0
1977	0
1978	0

hemoglobin A_2 and F. It is therefore probable that the father had toxic manifestations of benzene on a normal A_2 beta-thalassemia carrier state. The second patient with preleukemia had hypochromic microcytic anemia throughout his illness.[15,104,105] Serum iron was normal, but there was a high level of fetal hemoglobin (48%) associated with decreased hemoglobin A_2 (0.67%).[15,104,105] This patient was mentioned in Chapter 5. His family study was normal. It is well known that an increased level of fetal hemoglobin may occasionally be found in some cases of acute leukemia.[106] In addition to that, decrease or increase of hemoglobin A_2 and increase in hemoglobin F as a sign of inblanaced polypeptide chain synthesis have been found in some cases of acute leukemia.

In view of the above mentioned facts, we postulated that the refractory microcytic anemia in this patient with normal family studies was the result of a superimposed disturbance in the polypeptide chain synthesis, occuring in previously normal individuals.[104,105]

J. Cholera Vaccinations

Cholera vaccination is possibly a contributory factor or promoter in the development of leukemia with chronic benzene toxicity. As can be seen from Table 5, which shows the annual number of leukemic shoe workers in Istanbul during the period of 1967 to 1975, the peak incidence of leukemia due to chronic exposure to benzene was in 1973.[8,9,26] Furthermore, there was a marked increase in the annual number of the patients with acute leukemia associated with benzene exposure since 1971.[26] Although there were nine leukemic patients in the 4-year period (1967 to 1970), this number increased to 22 in the subsequent years (1971 to 1974). This sudden increase of leukemia after 1970 in shoe workers have led us to look for additional factors besides this chemical agent.[26]

With the appearance of some cholera cases in Istanbul at the beginning of September 1970, the majority of the population in Istanbul was vaccinated against this infectious disease. The cholera vaccination was performed in 14 of our leukemic patients associated with chronic benzene toxicity. However, we were not able to find out whether the remaining eight patients were vaccinated against cholera or not because during our study, this possibility was taken into consideration only after 1973. On the other hand, a variety of postvaccinal reactions are well known, such as the activation of inactive tuberculosis.

Therefore, in our cases of leukemia with chronic benzene toxicity, hypothetically, immunization to cholera may have played a contributory role, possibly as a promoting agent, in the development of leukemia, in addition to chronic exposure to benzene.[26] A study performed by Lenz and Pluznik[107] showed that clonal growth of murine granulocyte/macrophage progenitor cells (CFU$_c$) was inhibited by cholera toxin in soft agar cultures. This problem needs further study, particularly in animals.

K. The Mutagenic Effect of Benzene

Since recognition of benzene is hematotoxic in man and animals, evidences showing the mutagenic effect have been sought. As Kale and Baum[108] emphasized, although benzene induces chromosomal aberrations, it does not induce point mutations. Lyon,[109] Shahin,[110] and Brusick[111] reported that in vivo microbial assays carried out with or without the addition of microsomal homogenates of rat liver failed to give any indication of mutagenic activity of benzene.

In this study, Lyon[109] used sensitive plasmid containing strains of *Salmonella thyphimurium*. Benzene was tested in a plate incorporation assay both directly and in the presence of 59 microsomal fractions prepared from liver homogenates of rats. Benzene and microsomal fractions were used at concentrations of 0.1 to 1.0 $\mu\ell$, ranging from 1 to 50 $\mu\ell$/plate. Lyon[109] also performed a host-mediated assay in mice, some of which were treated with phenobarbital. The test organism was *S. thyphimurium* TA 1950, 0.1 mℓ of benzene was given to mice. There was no increase in mutation rate in benzene treated mice. Despite several modifications in experiments, the results of these in vitro assays were uniformly negative. Negative results were also obtained by Shahin and Brusick by similar trials.[110,111] Yet, benzene failed to induce gene mutations in the mouse lymphoma forward mutation assay and in somatic mutation test in *Drosophila*.[112,113] On the other hand, Kale and Baum have performed the sex-linked recessive lethal mutation test in *Drosophila* melanogaster males.[108,113] Benzene has not induced a significant number of mutations in this system. According to Kale and Baum,[108] this is in agreement with the results from *Salmonella*/microsome assay.

The cytogenic effect of benzene was tested by Hite et al.[114] by means of micronucleus test. This test is based on the detection of small chromatin particles in the cytoplasm of young erythrocytes from the bone marrow. In contrast to the trials mentioned above, the results showed a significant increase in the numbers of nuclei of polychromatic erythrocytes with micronuclei in the mice studied. Several compounds of varying chemical classes have been reported to do this in test animals, and many of these agents are to be carcinogenic.[114] Lyon[115] carried out the above mentioned test in rats; 0.025, 0.05, or 0.25 mℓ of benzene per kilogram of body weight was given to rats. The rats in the two higher dosage groups showed higher micronucleus counts than the corresponding controls.[115] Siou and Canan[116] performed a study of micronucleus test on mice, which were given 0.25 mℓ of benzene per kilogram of body weight. This trial showed that there was an increase in the number of cells with micronuclei.

Gad-El-Karim et al.[117] performed micronucleus test and metaphase analysis in mice which were treated with two doses of benzene (440/mg/kg) or toluene (800 to 1720 mg/kg). Toluene showed no clastogenic activity and reduced the clastogenic effect of benzene when mixture was given.[117] The pretreatment with some modifiers of mixed-function oxidase activity such as phenobarbital, 3-methylcholanthrelene, etc. did not protect against the clastogenic effects of benzene.

L. Cytogenetic Studies

Since the beginning of 1960, both human and other experimental studies performed by numerous investigators, particularly by Pollini and Colombi,[118,119] Vigliani and Forni,[120,121] Tough and Court-Brown,[122] showed that benzene exposure can induce a significant increase in the rate of numerical and structural chromosome abnormalities in somatic cells, both peripheral lymphocytes and bone marrow cells. Benzene-mediated effects on the genetic material of a cell, namely the cytogenetic effects, are either numerical changes in the number of chromosomes, or alterations in the structure of chromosomes. These chromosomal aberrations may be stable (C_s) or unstable (C_u). Because the changes induced by benzene on bone marrow are considered as similar to those induced by ionizing radiation, studies have been conducted to show whether benzene exposure might cause chromatic aberrations which occur as a result of the exposure to this chemical.

In 1964, Pollini and Colombi[118,119] described increased rates of aneuploid cells, up to 70%, some with structural aberrations in cultures of bone marrow and peripheral blood lymphocytes in workers with severe benzene-induced aplastic anemia. The abnormal metaphases were mainly hypodiploid, but some hyperdiploid cells were also found. Tough and Court-Brown[122] reported significantly increased rates of unstable chromosome (C_u cells) abberrations such as fragments, dicentric, and ring chromosome in 20 subjects exposed to benzene for 1 to 20 years in comparison with a group of controls. The workers studied had no signs of benzene-induced hematologic alterations at the time of cytogenetic examinations, although 14 of them had suffered from neutropenia 3 years earlier.

Numerous studies in benzene-induced hematologic disorders yielded similar changes. Vigliani and Forni, and Forni and associates[120,121,123-126] performed extensive cytogenetic studies on subjects with benzene hemopathy and individuals who had recovered from benzene toxicity. These studies showed that significantly increased numbers of both C_u and C_s cells were still present in most cases several years after cessation of exposure to benzene. They characterized the aberrations as unstable chromosomal changes (C_u) or stable chromosomal aberrations (C_s) such as deletions, translocation, inversion, and trisomies. The stable type was found to persist for several years following recovery from a hematologic disorder due to chronic benzene toxicity. In addition to these, aneuploidy or polyploidy have also been reported. Chromosomes of the C group seemed to be involved in stable changes with frequency higher than the expected.

In another study, Forni et al.[52] compared the chromosomal findings obtained in 34 workers, ten exposed to high concentrations of benzene, then to toluene, and 24 exposed only to toluene, with those obtained in 34 controls.[52] The proportion of unstable and stable chromosome changes were significantly higher in the benzene group as compared with the controls and toluene group, but not in the toluene group as compared with the controls. Similar structural and numerical chromosomal aberrations were also reported by Haberlandt and Mente[127] and Khan and Khan[128] in workers with chronic benzene toxicity, and by Erdoğan and Aksoy[129] in ten pancytopenic individuals with chronic benzene toxicity. In the Erdoğan and Aksoy series, four of whom were cytogenetically normal while in the remaining six, the following numerical and structural aberrations were found: aneuploidy, polyploidy in one and trisomy C in three, breakage, achromatic lesions, dicentric chromosomes, and endoreduplication in others (see Table 5 in Chapter 5). Also, in workers chronically exposed to levels of 5 to 25 ppm of benzene, despite the absence of hematological findings, there were significant increases in the frequency of chromosomal aberrations from blood lymphocytes.[122,130]

The cytogenetic study of Picciani[130] on 52 workers exposed to low levels of benzene, less than 10 ppm, has revealed an increase in aberration rates as compared to that of 44 controls. These aberrations were not related to the age of the workers. Picciani recommended that medical monitoring, including cytogenetic surveillance of industrial populations exposed to any carcinogenic agent, including benzene, is essential and that further research, with attempts to link cytogenetic analyses with better estimates of duration and extent of exposure to all constituents of the work environment, should be undertaken. Furthermore, Liniecki et al.[131] studied the chromosomes of lymphocytes of peripheral blood of 12 individuals who 8 to 12 years before, had suffered from leukopenia due to chronic benzene toxicity. Benzene exposure had ceased at the time of investigation, aneuploidy was found in 12.8%, in comparison with 6% in 16 controls.

Chromosome and chromatid aberrations in the euploid cells showed that the frequency of acentric fragments was higher in one-time patients than for the control groups. Furthermore, cytogenetic and cytokinetic changes were examined in Japan by Watanabe et al.[132] who used the differential staining method for sister chromatid in cultured lymphocytes from 16 female workers who had been ocupationally exposed to up to 40 ppm of benzene for 1 to 20 years and 7 controls. The evaluation of SCE (sister chromatid exchange) has been considered as a highly sensitive method for the estimation of mutagenic action of environmental chemicals. A survey was carried out about half a year after cessation of exposure to benzene. All the materials, such as paint materials, thinner, and varnish, contained no benzene. Only four among the workers showed mild hematological abnormalities, such as reduced hemoglobin and hematocrit or leukopenia, while no increase in the number of structural and numerical chromosome aberrations nor any change in cell-cycle kinetics were observed. A slight decrease in the frequency of SCE was detected. Funes-Cravioto et al.[133] have shown that organic solvent exposed technicians in chemical laboratories, and children of these technicians who had worked during pregnancy, had a significantly increased frequency of SCEs in cultured lymphocytes whereas no increase in SCEs frequency was found in the technicians who had past-exposure history to benzene and were currently exposed to toluene only.

On the other hand, Watanabe et al.,[132] in their study, observed a decrease in SCEs frequency in the workers half a year after the cessation of benzene use. According to the investigators, these findings may imply that at relatively low levels of certain chemicals, such as benzene or benzene epoxide, the chemicals mainly affect the naturally occurring DNA repair process at replication point, inhibiting the formation of the SCEs whereas at the higher concentration levels, the chemicals initially cause DNA damage, such as cross linking or base chemical adducts, thus necessitating the extra DNA repair and consequently increasing formation of the SCEs. On the other hand, a cytogenetic study was performed by Sarto et al.,[134] in 1984 in Italy.

There, 22 healthy workers engaged in a factory where benzene is produced from coal tar, and also toluene and xylene are manufactured, were subsequently exposed to low concentration of benzene, ranging from 0.2 to 12.4 ppm. Workers were divided into two groups according to the difference in the levels of exposure. Each exposed individual was paired with a suitable control group. Contrary to the results of Watanabe et al., there was no influence of benzene exposure on SCE level.

Despite this, when the overall group of exposed workers is compared to controls in relation to the frequencies of chromatid-type and chromosome-type aberrations, a statistically significant difference is observed only for the latter type of aberrations, and the increase of chromosome-type aberrations is significant when gaps are discarded. Chromosome-type aberrations were represented by breaks and acentric fragments. In the exposed group, two dicentric chromosomes were found, while none was found in controls. In Table 6, which is a slightly modified table of Sarto et al., the results of published cytogenetic studies of workers with chronic benzene toxicity are given.

M. Cytogenetic Findings in the Patients with Leukemia Due to Chronic Benzene Toxicity

There are some cytogenetic studies on cases of leukemia with chronic benzene toxicity. Sandberg[135] pointed out that the cytogenetic abnormalities reported in leukemia due to chronic exposure to benzene seem to indicate that these alterations may not be different from those found in other nonexposed cases of acute myeloproliferative or lymphoproliferative disorders.

Table 6
SUMMARY OF SOME CYTOGENETIC STUDIES ON WORKERS EXPOSED TO BENZENE

Reference	Subjects (no.)	Kind of exposure to benzene	Concentration (ppm)	Duration of exp. (years)	Years elapsed after the end of exp.	Hematological changes	Increase of SCA	Increase of SCE
Tough et al.[122]	20	Benzene as a solvent	25—150	1—7	2—3	+	+	n.d.
	12	Benzene as a solvent	25—150	1—23	1—6	n.s.	+	n.d.
	20	Refinery	12	2—26	Still working	n.s.	−	n.d.
Forni et al. 1971[125]	10	Retrogravure	131—532	1—22	n.s.	+[a]	+	n.d.
Forni et al. 1971[125]	32	Various processes	n.s.	n.s.	1—18[b]	+	+	n.d.
Haberlandt and Mente 1971[127]	12	Various processes	n.s.	n.s.	n.s.	n.s.	+	n.d.
Hartwich and Schwaniz 1972[128]	9	Refinery	25	3—7	Still working	+	+	n.d.
Erdoğan and Aksoy 1973[129]	10	Shoe workers	150—210	3—10	Still working	+	+	n.d.
Kahn and Kahn 1973[128]	7	n.s.	n.s.	11—20	Still working	+	+	n.d.
	8	n.s.	n.s.	2—5	Still working	+[c]	+	n.d.
Piccani 1979[130]	52	Industrial facilities	10	0.1—26	Still working	−	+	n.d.
Fredga et al. 1979	12	Industrial facilities	5—10	2—14	Still working	−	+	n.d.
Watanabe et al. 1980[132]	7	Ceramic decoration	2—50	2—12	0.5—1	+	−	Decrease
	9	Ceramic decoration	up to 9	1—20	0.5—1	+	−	Decrease
Sarto et al. 1984[134]	22	Distillation	2.2—12.8	3—35	Still working	−	+	n.d.

Note: Abbreviation, n.d. = not determined; n.s. = not specified; SCA = structural chromosomal aberrations; SCE = sister chromatid exchanges.

[a] Over ten subjects.
[b] Unknown for seven subjects.
[c] One out of 15 subjects.

Modified from Sarto, F., Cominato, I., Pinton, A. M., Brovedani, P. G., Merler, E., Peruzzi, M., Bianchi, V., and Levis, A. G., *Carcinogenesis*, 5, 827, 1984. With permission.

Table 7

THE RESULTS OF CYTOGENETIC STUDIES OBTAINED IN 10 LEUKEMIC PATIENTS WITH CHRONIC BENZENE EXPOSURE

Case no. and (age)	Types of leukemia	Duration of exp. (years)	Chromosomal aberrations	
			Numerical	Structural
1 (41)	Preleukemia	4/12	None	None
2 (21)	Preleukemia	5	Polyploidy 22%	Achromatic lesions 6%, acentric fragment 3%
3 (39)	Preleukemia	10	Trisomy C 22%	Breakage 7%, fusion and adhesion 2%
4 (36)	Preleukemia	7	None	Achromatic lesion 6%, dicentric chromosome 12%, breakage 4%, Ph_1 12%, thin elongated chromosome 8%
5 (24)	ALL	6	None	Achromatic lesion 5%, dicentric chromosome 3%
6 (43)	AML	6	Trisomy C 13%	Achromatic lesion 6%, dicentric chromosome 5%, secondary constriction
7 (42)	AML	10	Polyploidy	Achromatic lesion 4%, secondary constriction 3%
8 (48)	AMYL	3	Monosomy C 40%	Dicentric lesion 5%, pulverized figure 5%
9 (23)	A. erythroleukemia	11	None	Achromatic lesion 4%, dicentric chromosome 1%, breakage 1%, achromatic lesion 4%
10 (36)	A. erythroleukemia	15	None	Achromatic lesion 4%

From Erdoğan, G. and Aksoy, M., 3rd Meet. European African Div. ISH, *Abstracts*, 1, 8, 1975. With permission.

Forni and Moreo[124] have performed serial cytogenetic studies on a patient with acute myeloblastic leukemia who was exposed to benzene for 22 years. When the patient had aplastic anemia, the cultured blood cells showed a high incidence of stable and unstable chromosome changes. A few months later, cells with 47 chromosomes appeared in the bone marrow. When frank acute myeloblastic leukemia appeared, it was accompanied by the emergence of a 47 XX + C clone in the marrow and blood. In a case of benzene-induced erythroleukemia, the abnormal cells both of the erythroid and myeloid series had a pseudodiploid mode; 2 missing chromosomes of the E group with 2 small added markers (one ring chromosome and one minute centric chromosome).[136] As these abnormal mitoses probably occurred both in the erythroid and in the myeloid cells, it was suggested that the chromosome breaks and reconstructions had occurred in a common stem cell.

Sellyei and Keleman[137] found a 47 XY D clone in a 32-year-old male with subacute myeloid leukemia which developed 7 years following pancytopenia due to exposure to benzene. On the other hand, Hartwich et al.[138] observed only structural aberrations, such as increased rates of chromatid and chromosome breaks, in a case of benzene-induced acute myeloblastic leukemia.

Erdoğan and the present author[139] performed cytogenetic studies in 10 leukemic and 10 pancytopenic patients with chronic exposure to benzene. The results concerning the leukemic patients are summarized in Table 7. Numerical changes were found in five leukemic patients (polyploidy in two, trisomy C in two and monosomy C in one). Aberrations, such as acentric fragments, achromatic lesions, secondary constrictions, and dicentric breaks, were observed in nine patients. Among the findings observed in this study, the following are suggestive of leukemia: trisomy C, hypodiploidy (monosomy C), and the Ph chromosome (in a case of preleukemia). Interestingly, the Ph chromosome was observed by Erdoğan and the present author in a case of leukopenia due to chronic exposure to benzene for 4 years without the signs of leukemia.[139] After a 4-year period of nonexposure, these chromosomal aberrations disappeared. The other cytogenetic findings of these series are probably nonspecific. Table 5 (in Chapter 5) also show great variations in the severity of chromosomal aberrations obtained in ten patients with pancytopenia with chronic benzene toxicity.[139-141]

Rondanelli et al.[142] performed a cinomicrographic study in three patients with benzene erythropathy and in one with erythroleukemia due to occupational exposure to benzene showing changes which can be summarized as follows:

1. Morphological anomalies of erythroblasts in interphase: the interkinetic nuclei of erythroblasts may be atypical in form (irregular), volume (gigantism, or microerythroblasts), and structure (bi- and plurinucleated cells and structural abnormalities).
2. Morphological anomalies of erythroblasts in mitosis: (a) alteration of orientation and localization of chromosomes, particularly during metaphase, resembling metaphasic block. This phenomenon presents itself in the appearance of the ''star-metaphase'' or ''orientated altered metaphase''. This metaphasic block is similar to that produced by colchicine, and is therefore called c-mitosis; (b) chromosomal pathology in arrested mitosis has three manifestations: conglutination (distributed metaphases of chromatin masses or irregular form), fragmentation, and pyknosis.

New erythroblasts exposed in vitro to the action of benzene showed similar morphologic anomalies of interkinetic nuclei and of karyokinesis as described above.

Mittelman et al.[143] performed a cytogenetic study in two group patients with acute leukemia; 23 patients occupationally exposed to chemical solvents, insecticides, and petroleum products and 33 patients with no history of occupational exposure to potential mutagenic/carcinogenic agents. Acute myeloid leukemia was more common in the former, the acute monocytic variety in the latter. Both groups with abnormal metaphases had a poorer prognosis than those with normal metaphases. The detailed karyotypic findings showed striking differences between the two groups. In the nonexposed group, only 24.2% had chromosome aberrations compared with 82% of those exposed. In the exposed group, 84.2% of the patients had at least one of the four particular changes: monosomy 5 or 7, or trisomy 8 and 21, but in the nonexposed group, only one patient had monosomy 7 and one had trisomy 21. According to the investigators, these differences between the nonexposed and exposed group strongly indicate that the karyotypic patterns of the leukemic cells were in fact influenced by exposure.

Van der Berghe et al.[144] reported two patients with presumably benzene-induced malignant disorders. One of the patients had used solvent containing large quantities of benzene. The second patient apparently had been using glues containing benzene for several years. In both patients, initially preleukemia developed. The patients were cytogenitically monitored through the courses of their diseases and have been studied with the banding technique. In the first patient, in addition to familiar chromosome translocation, karyotypic abnormalities, such as acquired translocation and later, an almost complete exchange between the long arm of chromosome 15 and the short arm of chromosome 4, developed. The second patient had chromosome 7 monosomy. Later, the same anomaly was found in 100% of the bone marrow cells. Nearly 2 months later, all bone marrow cells were cytogenetically normal. With regard to the presence of cytogenetic findings following cessation of exposure to benzene, unfortunately there are very few reports on this problem.

Pollini et al.[145] performed cytogenetic studies in 4 individuals suffering from benzene hemopathy 10 years after exposure to this chemical. In three subjects a statistically significant presence of aneuploid cells was proved. Hyperdiploidy however, was not significant in any subject. Pollini et al. emphasized the nearly total disappearance of gross morphologic alteration of the karyotype in the subjects examined.

Furthermore, ten workers who were exposed to benzene following a spillage of about

1200 gallons of this chemical were cytogenetically studied by Clare[146] about 3 months later. Shortly afterwards, an assay of the urine of these individuals showed that substantial amounts of phenol were excreted. Cytogenetic study and analysis on SCE of these workers disclosed: all values were considered to be within a normal range. However, there were slightly more SCE in some exposed workers than the controls and there was a trend towards a positive association between the frequency of SCE recorded for each individual and the maximum value for the excretion of phenol in the urine on the day after the incident.

N. Experimental Studies

Kissling and Speck[147] and Moeschlin and Speck[148] studied the mechanism by which benzene affects bone marrow, using autoradiography techniques with ^{3}H-thymidine and ^{3}H-cytidine. These studies on rabbit bone marrow cells with benzene-induced pancytopenia revealed severely disturbed DNA and RNA synthesis and a high incidence of chromosome aberrations. As explained in Chapter 5, according to the investigators, this type of pancytopenia is primarily due to the interference of benzene with DNA and RNA synthesis and not to the failure of hematopoietic stem cells.[147,148] But as Freedman[149] pointed out rightly, these studies do not prove that the primary toxic event was indeed the inhibition of DNA synthesis, for these studies were performed after toxicity was manifest.

Furthermore, Koizumi et al.[150] have performed a study on the effects of benzene on DNA synthesis and chromosomes of cultured human lymphocytes and HeLa cells. Appropriate concentrations of benzene were added to human lymphocytes at the beginning of the 72-hr culture period. The combined incidence of chromatid gaps and breaks was significantly higher in cultures containing 1.1 or 2.2×10^{-3} M benzene than in untreated cultures. Koizumi et al.[150] attributed the decrease in tritium uptake in autoradiographs, prepared from both leukocytes and HeLa cells, to inhibition of DNA synthesis by benzene.

On the other hand, Morimoto[151] also obtained similar results, namely the increase in chromosome aberrations in cultured human leukocytes exposed to benzene. Morimoto and Koizumi[151-156] performed experiments on the chromosome mutagenicity of benzene, with particular reference to the combined effect with gamma radiation ($137C_s$) and the inhibitory effect on rejoining of radiation-induced chromosome breaks. Benzene induced mainly chromatin-type deletions, especially gaps, suggesting that the cells in their late S-G2 stage have a higher susceptibility to chromosome breakage by benzene.[151-156] On the other hand, the aberration yield dicentricity and rings induced by 100 rad can be significantly enhanced by treatment with benzene equal to or in excess of 0.2 mM.[155] These experiments suggest that the cytogenetic effects of benzene with radiation could be exclusively synergetic in aberrations, such as dicentricity and ring chromosomes, being additive in the other types of aberration.

In a similar study, phenol, the metabolite of benzene, was tested for its inhibitory effect on rejoining of radiation-induced chromosome breaks in cultured leukocytes by use of the dosage fractionation method.[156] Comparing the results obtained in these studies with those of the former, the inhibitory effect of phenol on rejoining of chromosome breaks was several tenfolds stronger than that of benzene.[155-156]

Furthermore, Philip and Jensen[157] demonstrated that chromatid deletions were present in metaphase chromosomes from bone marrow cells of rats analyzed 12 and 24 hr after subcutaneous injections of 2 mℓ of benzene per kilogram of weight. Furthermore, a chronic experiment with subcutaneous injections of benzene on rabbits showed an increase of structural chromosomal aberrations (gaps and breaks) and in a few instances, tetraploid mitoses were seen.[158] Tice et al.[159] exposed mice to 300 ppm benzene for 4 hr. This trial induced a significant increase in SCE of both sexes, but did not induce a significant increase in chromosomal aberrations of bone marrow cells of either sex, and inhibited cellular proliferation in bone marrow of male but not female mice.[159] Furthermore, the proliferative inhibition observed in male mice was greater 1 day after the exposure period. According to Tice et al.,[159] this observation suggests the delayed formation of toxic metabolites and/or delayed cellular reaction.

Styles and Richardson[160] exposed rats to benzene vapor at nominal concentrations in air at 1, 10, 100, and 1000 ppm acutely for 6 hr. Bone marrow cells from animals were examined for chromosomal abnormalities 24 hr after the end of the exposure period. The cytogenetic analysis showed a significant increase in the percentage of cells with chromosomal abnormalities, including or excluding gaps in animals exposed to 100 and 1000 ppm benzene. On the other hand, in the 10 and 1 ppm exposure groups, there were elevated levels of cells with abnormalities which showed evidence of being dose-related, although they were not statistically significant. According to investigators, at 10 and 1 ppm there was no significant difference in the incidence of aberrant cells in treated animals compared with controls, but the levels were, nonetheless, elevated and showed a dose response.[160]

In a study performed by Erexson et al.,[161] it was shown that mice exposed to benzene concentrations as low as 10 ppm for 6 hr disclosed an increased SCE frequency in peripheral blood B-lymphocytes and increased numbers of micronucleated polychromatic erythrocytes in their bone marrow. In another study, Erexson et al.[162] showed that benzene and its metabolites, such as phenol, catechol, 1,2,4-benzenetriol, hydroquinone, and 1,4-benzoquinone, induce SCEs in human T-lymphocytes from mononuclear leukocyte culture exposed in vitro without any additional activating system.

Pollini et al.[163] have studied by microcenstitometry the behavior of DNA of the interphasic nucleus in lymphocytes of four individuals 10 years after exposure to benzene. Despite the absence of cytogenetic findings, there was a statistically significant increase of Feulgen-negative material in all subjects. Pollini et al.[163] suggested the possibility of the DNA of the interphasic nucleus of the lymphocytes in benzene-poisoned subjects as being a "non-physiologic" condition, probably related to physico-chemical alteration of the DNA.

II. MALIGNANT LYMPHOMA AND BENZENE

There are few reports describing the relationship between occupational exposure to chemicals and risk of malignant lymphoma. Some occupations are most often considered to increase risk of Hodgkin's disease such as woodworking, carpentry, sawmill working, etc. According to Grufferman,[164] other occupational groups that have been considered as being at increased risk for Hodgkin's disease are schoolteachers, nurses, physicians and chemists. Furthermore, there were few case reports suggesting a possible etiological relationship beteen malignant lymphoma and exposure to benzene. As can be seen from Table 8, until 1974, only very few case reports appeared in literature. Bousser et al.[165] reported a case of lymphosarcoma which occurred 3.5 years after the cessation of benzene exposure. Since then, a few cases of malignant lymphoma associated with chronic exposure to benzene have also been described by some investigators as by Paterni and Sarnari[166] and Casirola and Santagari.[167] In 1974, present author and associates described six cases of Hodgkin's disease with a history of chronic exposure to benzene for periods ranging from 1 to 28 years with a mean of 11 years.[168] This series consisted of three shoe workers, one compass repairman, one car painter, and one a television repairman. Their ages varied between 27 and 61 with a mean of 39 years. These six patients belonged to a series of 94 cases of Hodgkin's disease of whom 25 were workers. All were studied in the Internal Clinic of Istanbul Medical School during the period of 1968 to 1972. Despite the lack of statistical data, we suggested that chronic exposure to benzene might play a role in the development of Hodgkin's disease. In addition, until 1985 we studied seven other cases in the intervening years: histiocytic lymphoma in a 42-year-old worker, lymphocytic lymphoma in a 47-year-old shoe worker, immunoangioblastic lymphadenopathy in a 35-year-old shoe worker, Hodgkin's disease in a 36-year-

Table 8
MALIGNANT LYMPHOMA AND CHRONIC BENZENE TOXICITY

Types of M.L.	No. of cases	Benzene exp. (ppm)	Comment	Ref.
Hodgkin's disease	1	—	Possible Hodgkin's disease diagnosed at necropsy	202
Lymphosarcoma	1	—	Benzene present in blood and tissues	165
Reticulum cell sarcoma	1	—	Benzene present in blood and tissues	166
Malign lymphoma	1	—	In 31-year-old male, 12 years previously he had benzene-induced anemia	167
Hodgkin's disease	6	150—210	Belonging to a series of 94 Hodgkin's disease cases in the period of 1968—1972	168
Hodgkin's disease	4	—	Diagnosed during the period of 1975—1985, all exposed to benzene	22
Histiocytic lymphoma	1	—	42-year-old worker with benzene exposure	22
Lymphocytic lymphoma	1	—	47-year-old shoe worker exposed to benzene	—
Immunoangioblastic lymphoma	1	—	34-year-old shoe worker exposed to benzene	—

old shoe worker, in an 45-year-old worker in a motor plant, and in a 31-year-old designer and a 30-year-old designer.[9,22,23] All were chronically exposed to benzene for 10 to 17 years.

There are numerous clinical and experimental data suggesting that benzene has a dele-terious effect on the lymphocytic or reticuloendothelial, as well as hemopotoietic, tissues. As early as in 1960, Wirtschafter and Bichel[169] showed that tissue responses could be observed in lymphnodes, spleen, thymus, and bone marrow of rats after a single injection of benzene. Reticuloendothelial cells of the macrophage type were increased in the lymph nodes and thymus following benzene injections, and returned to normal levels after a short time. Also, the spleen showed a marked proliferation of the reticuloendothelial cells of the pulp with morphological aberrations.

Goldwater[170] was the first to draw attention to lymphopenia in chronic benzene toxicity. This finding has generally been confirmed by several investigators. Advanced cases of both severe and mild chronic benzene toxicity will exhibit relative and absolute lymphopenia. These findings are discussed in Chapter 5. The present author and associates have found absolute lymphopenia in 24 out of 32 cases (75%) of aplastic anemia due to chronic exposure to benzene.[14,15,25] Thus, the absence of relative lymphocytosis in aplastic anemia with chronic benzene toxicity can be considered as a striking feature in the differentiation of this disorder from other types of aplastic anemia of idiopathic origin, or that it developed in connection with other hematotoxic agents. Furthermore, a study performed by Snyder et al.[171] in mice exposed to 300 ppm benzene vapor, clearly demonstrated that leukopenia was predominately due to lymphopenia. As we discussed in Chapter 5, a severe lymphocytosis is also to be observed in some cases of chronic benzene toxicity. Bernard[172] found three such cases, and three more with moderately increased lymphocyte count.

Considering the above mentioned facts, and that Hodgkin's disease may be a proliferative disorder of lymphoreticular tissue, the present author and associates[168] suggested the following:

1.   There is an etiological relationship between Hodgkin's disease and chronic benzene toxicity.
2.   Chronic exposure to benzene may act as a contributory factor in the development of Hodgkin's disease.

On the other hand, Vianna and Polan[173] performed a study of comparison on the 1950 to 1969 mortality rates for reticulum cell sarcoma, lymphosarcoma, and Hodgkin's disease among the males of the New York state residents, excluding New York City, aged 20 years or older in 14 different occupations with exposure to benzene or other coal-tar fractions, and among a nonexposed control group. Calculated relative risks for 14 occupations, based on crude death rates, were 1.6, 2.1, and 1.6 for reticulum cell sarcoma, lymphosarcoma, and Hodgkin's disease, respectively. Within the seven occupations for which age distribution were available, the observed number of deaths was significantly higher than the expected number for each type of lymphoma. Further analysis revealed that this excess was observed in men of 45 years of age or older. According Vianna and Polan,[173] these results are consistent with the possibility that chronic exposure to benzene and/or other coal derivates might be important in the etiology of this kind of tumour.

The study of Vianna and Polan was criticized by Enderline.[174] He accused the methodology deficiencies of this study. According to him, proportionate mortality analyses should have been used. When such analyses are done, no increased risk is observed.

As we mentioned earlier, in the studies on rubber workers with chemical exposures performed by Monson and Nakano[36] and later, by Monson and Fine[37] and McMichael et al.,[38-41] there was a slight increase in malignant lymphoma such as lymphosarcoma and reticulosarcoma. Recently, Norseth et al.[175] performed a study on cancer incidence among 2448 male rubber workers in the rubber industry in Norway. A suggested increased risk of bladder cancer, lymphoma, and leukemia was found in the footwear departments; the number of observed and expected cases was 4/2.81, 5/2.26, and 4/1.76, respectively. According to Norseth et al., a benzene-based glue was used in the footwear department up until 1940 and possibly in small amounts a few years thereafter. During the period of 1940 to 1974, a glue based on benzene (1%) with up to 4% aromatics, probably predominately benzene, was used.

There are also some studies showing a high rate of mortality from malignant lymphoma among pathologists, chemists, and persons who have handled chemicals containing benzene. Li and associates[176] reported in 1969 that chemists had an excess of cancer mortality and approximately half of the deaths were due to malignant lymphoma and cancer of pancreas. Similarly, Olin[177] in 1976, reported that Swedish chemistry graduates had a greater than expected cancer mortality with most of the excess attributed to leukemias and malignant lymphomas. In this study, the greatest excess was for Hodgkin's disease. Three patients were observed against 0.3 cases expected. According to Olin, most of the leukemia and lymphoma deaths were in organic chemists.[177] On the other hand, Olsson and Brandt,[178,179] in another study, reported that 12 patients with Hodgkin's disease (48%) were exposed to organic solvents. Contrary to this, only 6 from 50 referents (12%) were exposed to organic solvents.[179] According to Olsson and Brandt, it is possible that exposure to organic solvents may be relevant for the development of Hodgkin's disease, irrespective of the chemical configuration.

This study was criticized by Grufferman.[164] According to him, this case referent study was "crude" because exposure status was referred from occupations listed in medical records.

III. MULTIPLE MYELOMA AND CHRONIC BENZENE TOXICITY

In 1970, Torres et al.[180] reported two patients with IgG myeloma who were chronically exposed to benzene for 6 and 11 years. In our series of seven patients with multiple myeloma due to some etiologic factors, this hematologic malignancy was established due to benzene exposure in four.[9,22,26,181] Some clinical and laboratory data of these four patients with multiple myeloma are summarized in Table 9.

Table 9

SOME CLINICAL AND LABORATORY DATA IN FOUR PATIENTS WITH MULTIPLE MYELOMA ASSOCIATED WITH CHRONIC EXPOSURE TO BENZENE

Case no.	Age	Profession	Duration of exp.	Preceding pancytopenic period	Bone marrow cellularity (% of plasma cells)	Percentage plasma cells in blood film	Immunoglobulin	Bence-Jones proteinuria	Kidney involvement	Osteolitic lesions	Types of M.M.
1	36	Technician	7	+	Hypoplastic (5)	21[a]	IgG_{3400}	+ +	+	–	Bence-Jones
2	59	Manager in plastic plant	15	–	Normocellular (35)	0	IgG_{4888}	–	–	–	IgG, Lambda
3	63	Shoeworker	15	–	Normocellular (41)	0	—[b]	–	–	+ +	—
4	58	Furniture manufacturer	35	–	Normocellular (53)	0	IgA	–	–	+ + + +	IgA

[a] Before his death.
[b] Not performed.

From Aksoy, M., Erdem, Ş., Dinçol, G., Kutlar, A., Bakioğlu, I., and Hepyüksel, T., *Acta Haematol.*, 71, 116, 1984. With permission.

In only one of our patients, a technician in an airplane repair plant, there was a short period of moderate pancytopenia associated with hypoplastic bone marrow. A severe myeloma kidney and Bence Jones proteinuria developed rapidly. Despite the low percentage of atypical plasma cells in the bone marrow, 21% plasma cells were found in the peripheral blood smear just before death. The second case, a manager in a plastic plant, intermittently exposed to benzene for 15 years, had monoclonal hypergammaglobulinemia, IgG Lambda type, and 35% of atypical plasma cells in the bone marrow. Despite these findings, no sign of bone involvement appeared for a subsequent 10-year period. The third patient, a shoe-worker with 15 years of chronic exposure to benzene, had classical findings of multiple myeloma such as osteolytic lesions in cranium and vertebrae and 41% of myeloma cells in the bone marrow. The last patient, a furniture manufacturer was exposed to benzene nearly 35 years. There were numerous osteolytic lesions in the pelvic girdle and lower extremities. A pathologic fracture of femur developed. The serum contained IgA paraprotein. Bone marrow was normocellular with 53% plasma cells. In our series, the duration of exposure to benzene ranged between 7 and 35 years (mean 18 years). This period is longer than those found in leukemia and malignant lymphoma due to chronic benzene toxicity, 10.2 and 10.6 years, respectively.[9,22] Furthermore, the mean ages in these two series differed widely, 54 years in the former and 37.7 years in the latter. When one considers that multiple myeloma is usually encountered after 40 years of age, this result seems unsurprising.

Recently, De Couflé et al.[50] performed a historical cohort mortality study, comprising 259 employees of a chemical plant where benzene had been used in large quantities. According to the investigators, the findings are consistent with previous reports of leukemia following occupational exposure to benzene and raise the possibility that multiple myeloma could also be linked to benzene. Considering these findings, we are strongly inclined to accept that chronic benzene toxicity may cause multiple myeloma.

IV. LUNG CANCER AND CHRONIC BENZENE TOXICITY

As it is known, there is a clear etiologic relationship between the occurrence of lung cancer and exposure to some chemicals such as asbestos, nickel, etc. In 1976, considering the five lung cancers among the individuals associated with occupational exposure to benzene, the present author suggested a causal relationship between this chemical and lung cancer.[9,22,182] We have observed six cases of lung cancer among the individuals with chronic benzene toxicity.[9,22,182] In Table 10, some clinical and laboratory data are summarized.

In these cases, duration of exposure ranged between 8 and 35 years (mean: 17 years). Only three of these patients had mild hematologic abnormalities such as leukopenia or lymphocytopenia. In two patients, the histopathological diagnosis was oat-cell carcinoma (cases 1 and 2). On the other hand, all patients were heavy or moderate smokers. Because benzene available in Turkey did not contain benzo(a)pyrene, determined by ultraviolet spectrophotometry, lung cancer in these six patients cannot be attributed directly to this carcinogenic agent present in benzene.[15]

Benzene is mainly absorbed via the lungs and about 40% of the absorbed portion is exhaled unchanged; after 24 hr a small percentage of benzene can still be detected in the expired air.[183] Therefore, a carcinogenic effect of benzene in the lungs is possible. Because our patients were also smokers, a possible role of smoking in the development of the lung cancer should also be considered.

Recently, Delzell et al.[184] performed a case-control study among rubber workers. Results of matched analyses indicated that there was no association between lung cancer mortality and employment in rubber compounding, mixing, or curing jobs. Men who had worked at least 5 years in rubber repair operations, where there was potentially heavy exposure to particulates and fumes, experienced a twofold increase in lung cancer risk. According to

Table 10

SOME CLINICAL AND LABORATORY DATA IN SIX PATIENTS WITH LUNG CANCER ASSOCIATED WITH LONG-TERM EXPOSURE TO BENZENE

Case no.	Age	Profession	Duration of exp. (years)	Smoking habits, cigarettes per day	Diagnosis	No. of white blood cells ($10^9/\ell$)	Histopathology
1	31	Shoe worker	17	10/day for 10 years	Bronchogenic carcinoma of right upper lobe	4.9	Small anaplastic cell (oat cell)
2[a]	54	Physician	15	20/day for 40 years	Bronchogenic carcinoma of right upper lobe	7.3	Small anaplastic cell (oat cell)
3[b]	49	Orthopedic shoe manu-facturer	8	20/day for 10 years	Bronchogenic carcinoma of right middle lobe	3.0	Not performed
4	40	Shoe worker	15	40/day for 15 years	Bronchogenic carcinoma of right middle lobe	6.8	Undifferentiated epithelioma
5[c]	57	Dyer	35	20/day for 23 years	Bronchogenic carcinoma of left lower lobe	5.2	Undifferentiated carcinoma
6	40	Shoe worker	15	8—10/day for years	Right hilar carcinoma	8.4	Undifferentiated carcinoma (in cytologic examination of sputum)

[a] Worked in a laboratory of gynecological endocrinology.

[b] Died from liver and cerebral metastasis.

[c] Had liver metastasis and lymphocyte count 312/mm³, 6% lymphocytes in the differential count, possibly as the result of chronic benzene toxicity.

the investigators, there was a 70% excess of lung cancer risk among men employed in making special products, where the primary production activity was fuel cell manufacturing. The latter two findings were statistically of marginal significance.

V. PATHOGENESIS

The exact mechanism of the development of leukemia or related malignancies in benzene toxicity is not known. In this regard, there are some facts and hypotheses.

There are suggestions that benzene-induced leukemia results from a similar damage to bone marrow as that initiated by exposure to ionizing radiation, but the mechanism is not explained.[81] On the other hand, chromosomal changes (breaks and rearrangements) in human beings exposed to benzene, and chromosomal aberrations in animal experiments prompted Vigliani and Saita[29] and Forni and Vigliani[5] to speculate that chromosomal damage to bone marrow cells, giving rise to stable changes, might be responsible for the formation of abnormal cell clones, one of which might induce leukemia by selective advantage. However, a correlation between the duration and the degree of exposure and either the frequency of chromosome aberration or the degree of toxicity was not established. As it was pointed out, it is still doubtful whether chromosome changes are responsible for this malignant transformation in chronic benzene toxicity or not.[25]

According to the suggestions of some investigators, the carcinogenecity of aromatic hydrocarbons can be predicted from calculations of resonance energies of bonds in the K and L region on the basis of the localization theory of chemical reactions.[185-187] This theory predicts that compounds in which the K region is chemically reactive, but the L region is not, are carcinogenic. According to Sims et al.,[186] the activity of the K region resides largely in the ability to form epoxides which can be degraded or react with cellular constituents, e.g., proteins or nucleic acid. Despite the fact that the first product of benzene metabolism is an unstable compound called benzene oxide, the theoretical calculations predict that benzene is noncarcinogenic.[185-187]

As Snyder and Kocsis[185] rightly pointed out, the prediction of carcinogenecity based on calculations of resonance energy is subject to error. According to Snyder and Kocsis,[185,190] the study of the K region and K region epoxides may prove to be useful in the prediction of how benzene (or its epoxide) may react with nucleic acid either to depress the bone marrow, or produce leukemia. Snyder et al.[190] presented an example in clarifying their hypothesis of the main role of benzene epoxide exposure in the development of leukemia. The specific epoxide currently thought to induce carcinogenesis resulting from treatment with benzo(a)pyrene, for example, is a 7,8-diol-9,10-epoxide. It has been identified as a metabolite of benzo(a)pyrene which covalently binds to nucleic acids in the lung tissue, a major site of carcinogenesis by this compound, and it has been demonstrated to possess both mutagenic and carcinogenic properties.[190]

Since the reports of Hektoen,[191] it is known that benzene causes a deficiency in antibody production and a decrease in the phagocytic properties of leukocytes. The adverse effect of benzene on defense mechanisms has been demonstrated by markedly increased susceptibility to tuberculosis and pneumonia in rabbits repeatedly exposed to benzene vapor.[193,194] On the other hand, as we explained in Chapter 5, there are several studies showing important alteration or impairment of the lymphocytic system, serum immunoglobulins, and complements in chronic benzene toxicity. Recent observations suggest that the disturbed immune system may play an important role in the development of leukemia.[185,190] Therefore, the impaired immune mechanism may play a role in the development of malignancies, including leukemia, in chronic benzene toxicity.[195]

Another hypothesis is that of the activation of a latent leukemogenic virus or viral oncogens by chemical carcinogens including benzene. According to this "oncogene" theory of Heubner and Todero,[196] the genetic component of most vertabrates contains information for production of RNA-tumor viruses. Viral information, including the portion of the viral information responsible for inducing malignant transformation of the cell (oncogene), is inherited, but is normally repressed. Toxic insults, such as ionizing radiation and chemical carcinogens, may cause depression of oncogene.

Damage of hematopoietic stem cells by chronic benzene toxicity is responsible for the development of leukemia. As we explained in Chapter 5, there are several animal experiments which show clearly that benzene is responsible for a defect or absence in the hematopoietic stem cells.[197-199] Today, aplastic anemia, chronic and acute myeloid leukemia, preleukemia, paroxysmal nocturnal hemoglobinuria, myelofibrosis cyclic neutropenia, and constitutional anemias are considered disorders affecting hematopoietic stem cells.[200]

Any damage of stem cells can result in the disorders mentioned above. Recently, Stewart and Hess[201] have reported a young husband and wife simultaneously presenting aplastic anemia and acute myelomonocytic leukemia, respectively. No etiologic agent or exposure was identified. According to Stewart and Hess[201] a single causative agent might have induced stem cell changes in genetically dissimilar persons and produced a spectrum of pathologic features ranging from aplastic anemia to acute myelomonocytic leukemia. In fact, the present author and associates, and Vigliani and colleagues observed aplastic anemia and acute leukemia due to chronic benzene toxicity in the same workshops.[4,5,15] For the present author, the damage to hematopoietic stem cells, or an impairment in microenvironment caused by benzene exposure and individual genetic factors, might play the main role in the development of various hematological disorders observed in chronic benzene toxicity.

REFERENCES

1. **Santesson, G. G.,** Über Chronische Vergiftungen mit Steinkohlen Benzin vier Todesfalle, *Arch. Hyg.*, 31, 336, 1897.
2. **Le Noire, M. M. and Claude,** Sur un cas de purpura attribue a l'intoxication par le benzene, *Bull. Mem. Soc. Med. Hop. Paris*, 3, 1251, 1897.
3. **Delore, P. and Borgomano, J.,** Leucemie aique au cours de intoxication benzenique. Sur origine toxique de certaines leucemies aigues et leurs relations avec le anemies graves, *J. Med. Lyon*, 9, 227, 1928.
4. **Vigliani, E. C. and Forni, A.,** Benzene and leukemia, *Environ. Res.*, 11, 122, 1976.
5. **Forni, A. and Vigliani, E. C.,** Chemical leukemogenesis in man, *Ser. Haematol.*, 7, 211, 1974.
6. **Goguel, A., Cavigneaux, A., and Bernard, J.,** Le leucemies benzeniques de la region Parisienne 1950 et 1965 (Etudes de 50 observationes), *Nouv. Rev. Fr. Hematol.*, 7, 465, 1967.
7. **Tareef, E. M., Kontchaloskaya, N. M., and Zorina, L. A.,** Benzene leukemias *Acta Unio Int. Contra Cancrum*, 19, 751, 1963.
8. **Aksoy, M., Erdem, Ş., and Dinçol, G.,** Types of leukaemia in chronic benzene poisoning. A study in thirty-four patients, *Acta Haematol.*, 55, 65, 1976.
9. **Aksoy, M.,** Different types of malignancies due to occupational exposure to benzene: a review of recent observations in Turkey, *Environ. Res.*, 23, 181, 1980.
10. **Browning, E.,** *Toxicity and Metabolism of Industrial Solvents*, Elsevier, Amsterdam, 1965, 3.
11. DFG (Deutsche Forschungsgemeinschaft), *Benzene in the Work Environment*, Harald Boldt Verlag, Boppard, W. Germany, 1974.
12. **Lignac, G. O. E.,** Benzol Leukamie bei Menschen und Weissen Mausen, *Klin. Wochenschr.*, 12, 109, 1932.
13. **Aksoy, M.,** Leukemia in workers due to occupational exposure to benzene, *New Istanbul contrib. Clin. Sci.*, 12, 3, 1977.
14. **Aksoy, M., Dinçol, K., Erdem, Ş., Akgün, T., and Dinçol, G.,** Details of blood changes in 32 patients with pancytopenia associated with long-term exposure to benzene, *Br. J. Ind. Med.*, 29, 56, 1972.
15. **Aksoy, M. and Erdem, Ş.,** Follow-up study on the mortality and the development of leukemia in 44 pancytopenic patients with chronic exposure to benzene, *Blood*, 52, 285, 1978.

16. **Topuzoğlu, I.**, Reports on the solvents analysed in the laboratories of the Turkish Department of Labour (IŞGÜM), presented at the Meet. Comm. Benzene Problems in Turkey, Angora, May 1972.

17. **Aksoy, M., Erdem, Ş., and Dinçol, G.**, Leukemia in shoe-workers occupationally exposed to benzene, *Blood*, 44, 837, 1974.

18. **Goldstein, B. D.**, Clinical hematotoxicity of benzene, in *Carcinogenicity and Toxicity of Benzene. Advances in Modern Environmental Toxicology*, Vol. 4, Mehlman, M. A., Ed., Princeton Scientific Publishers, Princeton, N.J., 1983, 51.

19. **White, M. C., Infante, P. F., and Chu, K. C.**, A quantitative estimate of leukemia mortality associated with occupational exposure to benzene, *Risk. Anal.*, 2, 1195, 1982.

20. **Aksoy, M., Velicangil, S., Akgün, T., and Arat, M. A.**, Istanbulda benzenli materyel kullanan çeşidli iş-yerleri ve benzen üreten Ereğli, Üzülmez ve Karabük kok fabrikalarında benzen konsentrasyonu tayini üzerinde bir inceleme (Turkish), *Bull. Istanbul Med. Fac.*, 33, 129, 1970.

21. **Aksoy, M., Dinçol, K., Akgün, T., Erdem, Ş., and Dinçol, G.**, Haematological effects of chronic benzene poisoning in 217 workers, *Br. J. Ind. Med.*, 26, 296, 1971.

22. **Aksoy, M.**, Problems with benzene in Turkey, *Reg. Toxicol. Pharmacol.*, 1, 147, 1982.

23. **Aksoy, M., Özeriş, S., Sabuncu, H., Inaniçi, Y., and Yanardağ, R.**, A haemotological study on 231 workers exposed to benzene in Istanbul and Izmit during the period of 1983 and 1985, *Br. J. Ind. Med.*, in press.

24. **Aksoy, M.**, Chronic lymphoid leukaemia and hairy cell leukaemia due to chronic exposure to benzene. A report of three cases, *Br. J. Haematol.*, 67, 203, 1987.

25. **Aksoy, M.**, Benzene: leukemia and malignant lymphoma, in *Topics in Haematology*, Roath, S., Ed., Wright-PSG, Bristol, 1982, 105.

26. **Aksoy, M.**, Malignancies due to occupational exposure to benzene, *Am. J. Ind. Med.*, 7, 305, 1985.

27. **Penati, F. and Vigliani, E.**, Sub problema della mielopatic aplastifiche, pseudo aplastichemiche de benzols, *Rass. Med. Ind.*, 9, 345, 1938.

28. **Vigliani, E. C. and Saita, G.**, Leucemia emocitoblastica du benzolo, *Med. Lav.*, 34, 182, 1943.

29. **Vigliani, E. C. and Saita, G.**, Benzene and leukemia, *N. Engl. J. Med.*, 271, 872, 1964.

30. **Vigliani, E. C.**, Leukemia associated with benzene exposure, *Ann. N.Y. Acad. Sci.*, 271, 143, 1976.

31. **IARC Working Group**, IARC monographs on the evaluation of the carcinogenic risk of chemicals to humans, Vol. 29, 95, 1982.

32. **Infante, P. F., Wagoner, J. K., Rinsky, R. A., and Young, R. R. J.**, Leukaemia in benzene workers, *Lancet*, 2, 76, 1977.

33. **Ishimaru, T., Okeda, H., Tomiyasu, T., Tsuchimotu, T., Hoshino, T., and Ishimaru, M.**, Occupational factors in epidemiology of leukemia in Hiroshima and Nagasaki, *Am. J. Epidemiol.*, 93, 157, 1971.

34. **Committee on Toxicology, Assembly of Life Sciences, National Research Council**, *A Review of Health Effects of Benzene*, National Academy of Sciences, Washington, D.C., 1976.

35. **Vidanna, E. and Bross, I. D. J.**, Leukemia and occupations, *Prev. Med.*, 1, 513, 1972.

36. **Monson, R. and Nakano, K. K.**, Mortality among rubber workers. I. White male Union employees in Akron, Ohio, *Am. J. Epidemiol.*, 10, 289, 1976.

37. **Monson, R. and Fine, L.**, Cancer mortality and morbidity among rubber workers, *J. Natl. Cancer Inst.*, 61, 1047, 1978.

38. **McMichael, A., Spirates, R., and Kupper, C. C.**, An epidemiologic study of mortality within a cohort of rubber workers, *J. Occup. Med.*, 16, 458, 1974.

39. **McMichael, A. J., Spirates, R., Kupper, L. L., and Gamble, J. F.**, Solvent exposure and leukemia among rubber workers, *J. Occup. Med.*, 17, 234, 1975.

40. **McMichael, A. J., Spirates, R., Gamble, J. F., and Tousey, P. M.**, Mortality among rubber workers relationship to specific jobs, *J. Occup. Med.*, 18, 178, 1976.

41. **McMichael, A. J. Andelkovic, D. A., and Tyroler, H. A.**, Cancer mortality among rubber workers: an epidemiological study, *Ann. N.Y. Acad. Sci.*, 271, 125, 1976.

42. **Andelkovic, D., Taulbel, J., and Symons, M.**, Mortality experience of a cohort of rubber workers, *J. Occup. Med.*, 17, 389, 1976.

43. **Thorpe, J. J.**, Epidemiologic survey of leukemia in persons potentially exposed to benzene, *J. Occup. Med.*, 16, 375, 1976.

44. **Brown, S. M.**, Leukemia and potential benzene exposure, *J. Occup. Med.*, 17, 5, 1975.

45. **Baier, E. J. P.**, Statement to the Informal Hearing on the proposed OSHA benzene standard, U.S. Occupational Safety and Health Administration, Washington, D.C., July 1977.

46. **Brandt, L., Nilsson, P. G., and Mittelman, F.**, Occupational exposure to petroleum products in men with acute non-lymphocytic leukaemia, *Br. Med. J.*, 1, 553, 1978.

47. **Rinsky, R. A., Young, R. J., and Smith, A. B.**, Leukemia in benzene workers, *Am. J. Ind. Med.*, 2, 217, 1981.

48. **Tabershow, I. R.**, Protection from the adverse effects of benzene. I. Environmental cancer: a report to the public, Porter, E., Ed., *Tex. Rep. Biol. Med.*, 37, 162, 1978.

49. **Ott, M. G., Townsend, J. C., Fishback, W. A., and Langer, R. A.,** Mortality among individuals occupationally exposed to benzene, *Arch. Environ. Health,* 33, 3, 1978.
50. **De Couflé, P., Blattner, W. A., and Blair, A.,** Mortality among chemical workers exposed to benzene and other agents, *Environ. Res.,* 30, 16, 1983.
51. **Girard, R. and Revol, L.,** La frequence d'une exposition benzenique au cours de hemopathies graves, *Nouv. Rev. Fr. Hematol.,* 10, 477, 1970.
52. **Forni, A., Pacifico, E., and Limonte, A.,** Chromosome studies in workers exposed to benzene or toluene or both, *Arch. Environ. Health,* 22, 273, 1971.
53. **Arp, E. W., Wolf, P. H., and Checkoway, H.,** Lymphocytic leukemia and exposures to benzene and other solvents in the rubber industry, *J. Occup. Med.,* 25, 598, 1983.
54. **Chekoway, H., Wilcosky, T., Wolf, P., and Tyroler, H.,** An evaluation of the associations of leukemia and rubber industry solvent exposures, *Am. J. Ind. Med.,* 5, 239, 1984.
54a. **Yin, S.-N., Li, G. L., Tain, F. D., Fu, Z. I., Jin, C., Chen, Y. Z., Luo, S. J., Ye, P. Z., Zhang, J. Z., Wang, G. C., Zhang, X. C., Wu, H. N., and Zhong, Q. C.,** Leukaemia in benzene workers: a retrospective cohort study, *Br. J. Ind. Med.,* 44, 124, 1987.
55. **Bizzozero, O. J., Jr., Johnson, K. G., and Ciocco, A.,** Radiation related leukemia in Hiroshima and Nagasaki. I. Distribution, incidence and appearance time, *N. Engl. J. Med.,* 274, 1095, 1966.
56. **Hochino, T.,** Etiologic role of ionizing radiation on the development of leukemia: clinical and statistical studies on leukemia in patients treated with radio-iodine and in atomic bomb survivors, *Acta Haematol. (Japan),* 31, 825, 1968.
57. **Lange, R. D., Molony, W. C., and Yamaki, T.,** Leukemia in atomic bomb survivors. General observations, *Blood,* 9, 574, 1954.
58. **Aksoy, M., Erdem, Ş., Dinçol, G., Bakioğlu, I., and Kutlar, A.,** Aplastic anemia due to chemicals and drugs: a study of 108 patients, *Sex. Trans. Dis. Suppl.,* 11, 347, 1984.
59. **Court-Brown, W. M. and Doll, R.,** Leukaemia and aplastic anaemia in patients irradiated for ankylosing spondylitis, Special Report Series No. 295, p. 218, London, MRC.
60. **Court-Brown, W. M. and Doll, R.,** Adult leukaemia: incidence in mortality in relation to aetiology, *Br. Med. J.,* 1, 1065, 1959.
61. **Abbat, J. D., Forran, E. A., and Grene, R.,** AML after radioactive iodine (131) therapy, *Lancet,* 1, 782, 1956.
62. **Aksoy, M.,** Chronic myeloid leukemia after radioiodine therapy for hyperthyroidism, *New Istanbul Contrib. Clin. Sci.,* 12, 185, 1973.
63. **Aksoy, M., Dinçol, K., and Erdem, Ş.,** Acute leukemia due to chronic exposure to benzene, *Am. J. Med.,* 52, 160, 1972.
64. **Ruscetti, F., Collins, S., Gallaghar, R., and Gallo, R.,** Inhibition of human myeloid differentiation in vitro by Primate type C-Retro virus, in *Modern Trends in Human Leukemia. III.,* Neth, R., Gallo, R., Mannweiler, K., and Moloney, W. C., Eds., Springer-Verlag, Basel, 1979.
65. **Duvoire, M., Derobert, C., and Albahry, C.,** Examen pathologique d'une cas d'anemia mortelle survives vingt mois la cessation du travail dans benzol, *Sang,* 15, 356, 1942.
66. **Hernberg, S., Savilahti, M., Ahlman, K., and Asp, S.,** Prognostic aspects of benzene poisoning, *Br. J. Ind. Med.,* 23, 204, 1966.
67. **Di Guglielmo, G. and Iannacone, A.,** Inhibition of mitosis and recessive changes of erythroblasts in acute erythropathy caused by occupational benzene poisoning, *Acta Haematol.,* 19, 144, 1958.
68. **Sakol, M. J.,** Testimony at OSHA Hearing on benzene, U.S. Occupational Safety and Health Administration, Washington, D.C., 1977.
69. **Markham, R. E., Shih, T., and Rowley, P. T.,** Erythroleukemia manifesting δβ-thalassemia, *Hemoglobin,* 7, 71, 1983.
70. **Marchand, M.,** Un cas mortel de leucose benzolique, *Arch. Mal. Prof. Hyg. Toxicol. Ind.,* 21, 576, 1958.
71. **Wintrobe, M. M., Lee, G. R., Boggs, Bethell, T. C., Foerster, J., Athens, J. W., and Lukens, J. N.,** *Clinical Hematology,* 8th ed., Lea & Febiger, Philadelphia, 1981, 700.
72. **Emil-Weil, P.,** Le leucemie post benzolique, *Bull. Mem. Soc. Med. Hop. Paris,* 47, 193, 1932.
73. **Kahler, H. J. and Merker, H.,** Chronische myeloische Leukemia nach langjährigem Kontakt mit Benzol, *Dtsch. Med. Wochenschr.,* 86, 135, 1961.
74. **Marchal, G. and Duhamel, G.,** L'anemie hemolytique dans le leucemies, *Sang,* 21, 254, 1960.
75. **Hellriegel, K. P., Fohlmeister, I., and Schaefer, H. E.,** Aplastic anemia terminating in leukemia, in *Aplastic Anemia. Pathophysiology and Approaches to Therapy,* Heimpel, H., Gordonsmith, E. C., Heit, W., and Kubanek, B., Eds., Springer-Verlag, Berlin, 1979, 47.
76. **Falconer, E. H.,** Instance of lymphocytic leukemia following benzol poisoning, *Am. J. Med. Sci.,* 186, 353, 1933.
77. **Aksoy, M.,** Benzene as a leukemogenic and carcinogenic agent, *Am. J. Ind. Med.,* 8, 9, 1985.
78. **Di Guglielmo, G. and Ricci, M.,** Prima descrizione un caso di mielo eritremic da benzolo nella sua varieta cronica, *Settim. Med.,* 46, 365, 1958.

79. **Daily, S., Millet, B. and Gaultier, C.,** Leucemie tricholeucocytes chez une sujet expose du benzene, *Arch. Mal. Prof. Hyg. Toxicol. Ind.,* 39, 487, 1979.

80. **Tangün, Y.,** personal communication.

81. **Cronkite, E. P.,** Evidence for radiation and chemicals as leukemogenic agents, *Arch. Environ. Health,* 3, 297, 1961.

82. **De Gowin, R. L.,** Benzene exposure and aplastic anemia followed by leukemia 15 years later, *JAMA,* 185, 748, 1963.

83. **Goldstein, B. D.,** Hematotoxicity in humans, in *Benzene Toxicity. A Critical Evaluation,* Laskin, S. and Goldstein, B. D., Eds., McGraw-Hill, New York,1978; *J. Toxicol. Environ. Health Suppl.,* 2, 69, 1977.

84. **Kinoshita, Y., Terade, H., Saito, H., et al.,** A case of myelogeneous leukemia, *J. Jpn. Haematol. Soc.,* 85; as cited in **Goldstein, B. D.,** *Benzene Toxicity: A critical Evaluation,* Laskin, S. and Goldstein, B. D., Eds., McGraw-Hill, New York 1978; *J. Toxicol. Environ. Health Suppl.,* 2, 69, 1977.

85. **Erf, L. A. and Rhodas, C.,** The hematological effects of benzene (benzol) poisoning. I, *J. Ind. Hyg. Toxicol.,* 21, 421, 1939.

86. **Pugni, M., Sinibaldi, V., and Galli, F.,** Evaluzione leucosica acute di caso di mielopatia involutiva benzolica, *Haematologica,* 56, 57, 1971.

87. **Bernard, J. and Braier, L.,** Le leucoses benzinique, in *Proc. 3rd Int. Congr. Int. Soc. Hematology,* Grune & Stratton, New York, 1951, 251.

88. **Barber, R. and Spiers, P.,** Oxford survey of childhood cancer. Progress Report II, *Mon. Bull. Minist. Health Public Health Lab. Serv.,* 23, 46, 1964.

89. **Gunz, F. W.,** Problems in leukemia etiology, *Plenary Sessions Scientific Conference,* Proc. 13th Int. Congr. Hematology, Lehmann, Munich, 1970, 48.

90. **Undritz, E. and Schniyder, F.,** Vier Geschwister mit chronischer lymphatischer Leukemieim Wallis, *Schweiz. Med. Wochenschr.,* 101, 1779, 1971.

91. **Li, F. P., Marchetto, D. J., and Vawter, G. F.,** Acute leukemia and preleukemia in eight males in a family an x-linked disorder, *Am. J. Hematol.,* 6, 61, 1979.

92. **MacMahon, B. and Levy, M. A.,** Prenatal origin of childhood leukemia evidence from twins, *N. Engl. J. Med.,* 270, 1082, 1964.

93. **Wald, N., Borges, W. H., Li, C. C., Turner, J. H., and Harnois, M. S.,** Leukaemia associated with mongolism, *Lancet,* 1, 1228, 1961.

94. **Bloom, G. E., Warner, S., Gerald, P. S., and Diamond, L. K.,** Chromosome abnormalities in consti-tutional aplastic anemia, *N. Engl. J. Med.,* 274, 8, 1966.

95. **Sawitsky, A., Bloom, D., and German, J.,** Chromosomal breakage and acute leukemia in congenital telangiectasic erythema and shunted growth, *Ann. Intern. Med.,* 65, 487, 1966.

96. **Hecht, R., Koler, R. D., Rigas, D. A., Dohnke, G. S., Case, M. P., Tisdale, V., and Miller, R. W.,** Leukaemia and lymphocytes in ataxia teleangiectasia, *Lancet,* 2, 1193, 1966.

97. **Gallinelli, R. and Traldi, A.,** L'emopatia benzenica, tre casi benzolisma cronico di cui duie mortali leukemie acute (Panmyelofitisi acute), *Med. Lav.,* 54, 169, 1963.

98. **Hartwich, G. and Schwanitz, G.,** Chromosome Aberrationen bei einer Benzol Leukamia, *Dtsch. Med. Wochenschr.,* 97, 1228, 1969.

99. **Hartwich, G. and Schwanitz, G.,** Chromosome Untersuchungen mach chronischer Benzol Exposition, *Dtsch. Med. Wochenschr.,* 97, 45, 1972.

100. **Stieglitz, R., Stobbe, H., and Schüttermann, W.,** Leukosen nach Benzol, *Arch. Geschwulstforsch.,* 44, 145, 1974.

101. **Tabershaw, I. R. and Lamm, S. H.,** Benzene and Leukemia, *Lancet,* 2, 867, 1977.

102. **Tabershaw, I. R.,** Protection from the adverse effects of benzene. I. Environmental cancer: a report to the public, Porter, E., Ed., *Tex. Rep. Biol. Med.,* 37, 162, 1978.

103. **Infante, P. F.,** Protection from the adverse effects of benzene. I. Environmental cancer: a report to the public, Porter, E., Ed., *Tex. Rep. Biol. Med.,* 37, 62, 1978.

104. **Aksoy, M., Erdem, Ş., and Dinçol, G.,** The reaction of normal and thalassemia individuals to benzene poisoning the diagnostic significance of such studies, in *Abnormal Haemoglobins and Thalassaemia Di-agnostic Aspects,* Schmidt, R. M., Ed., Academic Press, New York, 1975, 267.

105. **Aksoy, M., Erdem, Ş., Schroeder, W. A., and Huisman, T. H. J.,** Hemoglobins A_2 and F in chronic benzene poisoning, in Int. Istanbul Symp. Abnormal Hemoglobins and Thalassemia, Aksoy, M., Ed., Scientific and Technical Research Council of Turkey (TBTAK), Angora, 1975, 197.

106. **Weatherall, D. J. and Clegg, J. B.,** *The Thalassemia Syndromes,* 3rd ed., Blackwell Scientific, Oxford, 1981.

107. **Lenz, R. and Pluznik, D. H.,** Inhibition by cholera toxin of clonal growth of murine in soft agar cultures, *Exp. Hematol.,* 10. 620, 1982.

108. **Kale, P. G. and Baum, J. W.,** Genetic effects of benzene in drosophila melanogaster males, *Environ. Res.,* 5, 223, 1983.

109. **Lyon, J. P.**, Mutagenecity studies with benzene, *Diss. Abstr. B*, 36, 55, 1976.
110. **Shahin, M. M.**, Unpublished results, University of Alberta, Edmonton, Canada, 1977; as cited in Dean, B. J., Genetic toxicology of benzene, toulene, xylenes and phenols, *Mutat. Res.*, 47, 75, 1978.
111. **Brusick, D. J.**, Mutagenecity of benzene, Litton Bionetics Report, March 25, 1977; as cited in **Raalte, H. G. S.**, *Evaluation of Benzene Toxicity in Man and Animals*, DGMK Berichte, Res. Rep. 174-6, Druckerei Hermann Lange, Hamburg, 1980.
112. **Libowitz, H., Brusick, D., Matheson, D., Reed, M., Goode, S., and Roy, G.**, The genetic activity of benzene in various short term in vitro and in vivo assays for mutagenecity, Presented at the Environmental Mutagen Soc. Meet., Colorado Springs; as cited in **Kale, P. G. and Baum, J. W.**, *Environ. Res.*, 5, 223, 1983.
113. **Nylander, P., Olafsson, H., Rasmuson, B., and Svahlin, H.**, Mutogenic effects of petrol in drosophila melanogaster. I. Effects of benzene and 1,2-dichloroethane, *Mutat. Res.*, 57, 163, 1978.
114. **Hite, M., Pecharo, M., Smith, I., and Thornton, S.**, The effect of benzene in the micronucleus test, *Mutat. Res.*, 77, 149, 1980.
115. **Lyon, J. P.**, Mutagenecity Studies with Benzene, Ph.D. thesis, University of California, Berkeley, 1975; as cited in **Dean, B. J.**, *Mutat. Res.*, 47, 75, 1978.
116. **Siou, G. and Canan, L.**, Activite mutagene du benzene et du benzo (a) Pyrene. Mise en evidence par la technique des corps de Howell-Joly (micronucleus test), Cah de notes document, INRSN.
117. **Gad-El-Karim, M. M., Harper, B. L., and Legator, M. S.**, Modifications in the myeloclastogenic effect of benzene in mice with toulene, phenobarbital, 3-methylcholanthrene, Aroclor,1254 and SKF-525A, *Mutat. Res.*, 135, 225, 1984.
118. **Pollini, G. and Colombi, R.**, I l danno cromosamico medollare nell' anemia aplastica benzolica, *Med. Lav.*, 55, 241, 1964.
119. **Pollini, G. and Colombi, R.**, I l danno cromosomico dei linfociti nell' emopatia benzenica, *Med. Lav.*, 55, 641, 1964.
120. **Vigliani, E. C. and Forni, A.**, J. Occup. Med. (Milan) letter, *J. Occup. Med.*, 11, 148, 1969.
121. **Vigliani, E. C. and Forni, A.**, Leucomogenesis professionale, *Minerva Med.*, 57, 3952, 1964.
122. **Tough, I. M. and Court-Brown, W. M.**, Chromosome aberrations and exposure to ambient benzene, *Lancet*, 1, 684, 1965.
123. **Forni, A.**, Chromosome changes and benzene exposure. A review, *Rev. Environ. Health*, 3, 5, 1979.
124. **Forni, A. and Moreo, L.**, Cytogenetic studies in a case of benzene leukaemia, *Eur. J. Cancer*, 3, 251, 1967.
125. **Forni, A., Ceppelini, E., Pacifico, E., Vigliani, E. L.**, Chromosome changes and their evaluation in subjects with post exposure to benzene, *Arch. Environ. Health*, 23, 385, 1971.
126. **Vigliani, E. C. and Forni, A.**, Leucomogenesis professionale, *Minerva Med.*, 3952, 1964.
127. **Haberlandt, W. and Mente, B.**, Aberration an der Chromosomzahl und Strüktur bei Benzol exponierten Industriearbeitern, *Zentralbl. Arbeitsmed. Arbeitsschutz*, 21, 338, 1971.
128. **Khan, H. and Khan, M. H.**, Cytogenetich Untersuchungen bei chronischer Benzol-Exposition, *Arch. Toxicol.*, 31, 39, 1973.
129. **Erdoğan, G. and Aksoy, M.**, Cytogenetic studies in 13 patients with pancytopenia and leukaemia associated with long-term exposure to benzene, *New Istanbul Contrib. Clin. Sci.*, 10, 230, 1973.
130. **Picciani, D.**, Cytogenetic study of workers exposed to benzene, *Environ. Res.*, 19, 33, 1979.
131. **Liniecki, J., Bajerska, A., and Gluszcowa, M.**, Analiza kariologiczna limfocytow krwi obwodowey u osob z przebytym przewleklym zatruciem benzenem, *Med. Pr.*, 22, 187, 1971; as cited in **Raalte, H. G.**, *Evaluation of Benzene Toxicity in Man and Animals*, DGMK Berichte, Res. Rep. 174-6, Druckerei Hermann Lange, Hamburg, 1980, 98.
132. **Watanabe, T., Endo, A., Kato, Y., Shima, S., Watanabe, T., and Ikeda, M.**, Cytogenetics and cytokinetics of cultured lymphocytes from benzene exposed workers, *Int. Arch. Occup. Environ. Health*, 46, 31, 1980.
133. **Funes-Cravioto, F., Zapata-Grayon, C., Kolmodin-Hedman, B., Lambert, B., Linsten, J., Norberg, E., Nordenskyold, M., Olin R., and Swendsson, A.**, Chromosome aberrations and sister chromatid exchange in workers in chemical laboratories and a roto printing factory and in children of women and laboratory workers, *Lancet*, 2, 322, 1977.
134. **Sarto, F., Cominato, I., Pinton, A. M., Brovedani, P. G., Merler, E., Peruzzi, M., Bianchi, V., and Levis, A. G.**, A cytogenic study on workers exposed to low concentrations of benzene, *Carcinogenesis*, 5, 827, 1984.
135. **Sandberg, A. H.**, The chromosome in human cancer and leukemia, American Elsevier, New York, 1979.
136. **Forni, A. and Moreo, L.**, Chromosome studies in a case of benzene induced erythroleukaemia, *Eur. J. Cancer*, 5, 459, 1969.
137. **Sellyei, L. and Keleman, E.**, Chromosome study in a case of granulocytic leukemia with pelgerisation seven years after benzene pancytopenia, *Eur. J. Cancer*, 7, 83, 1971.

138. **Hartwich, G., Schwanitz, G., and Becker, J.,** Chromosome anomalies in a case of benzene leukemia, *Dtsch. Med. J.,* 14, 449, 1969.

139. **Erdoğan, G. and Aksoy, M.,** Cytogenetic studies in 20 patients with pancytopenia and leukaemia with long-term exposure to benzene, European and African Division Int. Soc. Hematology, 3rd Meeting, London, August 24—28, *Abstracts 1,* 6, 11, 1975.

140. **Aksoy, M., Erdem, Ş, Erdoğan, G., and Dinçol, G.,** Acute leukaemia in two generations following chronic eposure to benzene, *Hum. Hered.,* 24, 70, 1974.

141. **Aksoy, M., Erdem, Ş., Erdoğan, G. and Dinçol, G.,** Combination of genetic factors to benzene in the aetiology of leukaemia, *Hum. Hered.,* 26, 149, 1976.

142. **Rondanelli, E. G., Gerna, G., and Maglinli, E.,** Pathology of erythroblastic mitosis in occupational benzene erythropathy and erythremia, in *Bibliothica Haematologica,* No. 35, Rondanelli, E. G., Ed., S. Karger, Basel, 1970.

143. **Mittelman, F., Brandt, L., and Nilsson, P.,** Relation among occupational exposure to potential mutagenic/ carcinogenic agents, clinical findings and bone marrow chromosomes in acute non-lymphocytic leukemia, *Blood,* 52, 1229, 1978.

144. **Van Der Berghe, A., Louwagie, A., Broeckaert-Van Orshoven, G. D., and Verwilghen, R.,** Chromosome analysis in two unusual malignant disorders presumably induced by benzene, *Blood,* 53, 558, 1979.

145. **Pollini, G. and Biscaldi, G. P.,** Persistence of karyotype alterations in lymphocytes ten years after benzene poisoning, *Med. Lav.,* 67, 465, 1976; (English Abstract) *Ind. Hyg. Dig.,* 41, 44, 1977.

146. **Clare, M. G.,** Chromosome analysis from peripheral blood lymphocytes of workers after an acute exposure to benzene, *Br. J. Ind. Med.,* 41, 249, 1984.

147. **Kissling, M. and Speck, B.,** Further studies on experimental benzene induced aplastic anemia, *Blut,* 25, 97, 1972.

148. **Moeschlin, S. and Speck, B.,** Experimental studies on the mechanism of action on benzene on the bone marrow (radiographic studies using ^{3}H-thymidine), *Acta Haematol.,* 38, 1041, 1967.

149. **Freedman, M. L.,** The molecular site of benzene toxicity, in Benzene Toxicity: A critical Evaluation, Laskin S., and Goldstein, J. B., McGraw-Hill, New York, 1978; *J. Toxicol. Environ. Health Suppl.,* 2, 37, 1977.

150. **Koizumi, A., Dobeshi, Y., Tachibina, Y., Tsuda, K., and Katsumuma, H.,** Cytokinetic and cytogenetic changes in cultures of human leucocytes and Hela cells induced by benzene, *Ind. Health (Japan),* 12, 23, 1974; as cited in **Dean, B. J.,** *Mutat. Res.,* 47, 45, 1978.

151. **Morimoto, K.,** Combined cytogenetic effects of benzene and radiation on cultured human lymphocytes, *Jpn. J. Ind. Health,* 17, 106, 1975; as cited in **Dean, B. J.,** *Mutat. Res.,* 47, 75, 1978.

152. **Morimoto, K. and Koizumi, A.,** Inhibition of rejoining of radiation induced chromosome lesions and induction of sister chromatide exchanges: effects of benzene or its metabolites in cultured human leucocytes (meeting abstract), *Jpn. J. Hum. Genet.,* 23, 279, 1978.

153. **Morimoto, K., Koizumi, Tachibana, Y., and Dubeshi, Y.,** Inhibition of repair of radiation induced chromosome breaks: effect of phenol in cultured human leucocytes, *Jpn. J. Ind. Health,* 18, 478, 1976.

154. **Morimoto, K.,** Combined cytogenetic effects of benzene and radiation cultured human lymphocytes, *Jpn. J. Ind. Health,* 17, 100, 1975.

155. **Morimoto, K.,** Inhibition of repair of radiation-induced chromosome breaks, *Jpn. J. Ind. Health,* 17, 166, 1975.

156. **Koizumi, A., Dobashi, Y., and Tachibana, Y.,** Chromosome changes induced by industrial chemicals, *Jpn. J. Ind. Health,* 21, 3, 1979.

157. **Philip, P. and Jensen, M. K.,** Benzene induced chromosome abnormalities in rat bone marrow cells, *Acta Pathol. Microbiol. Scand.,* 78, 489, 1970.

158. **Kissling, M. and Speck, B.,** Chromosome aberration in experimental benzene intoxication, *Helv. Med. Acta,* 36, 59, 1972.

159. **Tice, R. R., Vogt, T. F., and Costa, D. L.,** Cytogenetic effects of inhaled benzene in murine bone marrow, in *Genotoxic Effects of Airborne Agents,* Tice, R. R., Costa, D. L., and Schaich, K. M., Plenum Press, New York, 1982, 257.

160. **Styles, J. A. and Richardson, C. R.,** Cytogenetic effects of benzene: dosimetric studies on rats exposed to benzene vapour. *Mutat. Res.,* 135, 203, 1984.

161. **Erexson, G. L., Wilmer, J. L., and Kligerman, A. D.,** Induction of sister chromatid exchanges and micronuclear in male DBAI mice after inhalation of benzene, *Environ. Mutagen.,* 6, 408, 1984.

162. **Erexson, G. L., Wilmer, J. L., and Kligerman, A. D.,** sister chromatid exchange induction in human lymphocytes exposed to benzene and its metabolites in vitro, *Cancer Res.* 45, 2477, 1985.

163. **Pollini, G. et al.,** Increase of Feulgen-reactive material in peripheral blood lymphocytes of benzene poisoned subjects after ten years of intoxication. Preliminary data, *Med. Lav.,* 67(Suppl. 5), 506, 1976; English abstract, *Ind. Hyg. Dig.,* 43, 1977.

164. **Grufferman, S.,** Hodgkin's disease, in *Cancer Epidemiology and Prevention.* Schottenfeld, D. and Fraumeni, J. F., Jr., W. B. Saunders, Philadelphia, 1982, 739.

165. **Bousser, J. R., Neyde, R., and Fabre, A.,** Un cas hemopathie benzolique tres returdée a type lymphosarcome, *Bull. Mem. Soc. Med. Hop. Paris,* 63, 1100, 1947.

166. **Paterni, L. and Sarnari, V.,** Involutional myelopathy due to benzene poisoning, with appearance of a mediastinal reticulosarcoma at an advanced stage, *Securitas,* 50, 55, 1965.

167. **Casirola, G. and Santagari, G.,** Considerazioni sue due casi climento primente di lin forni a sede Splenica, *Haematologica,* 54, 85, 1969.

168. **Aksoy, M., Erdem, Ş., Dinçol, K., Hepyüksel, T., and Dinçol, G.,** Chronic exposure to benzene as a possible contributory etiologic factor in Hodgkin's disease, *Blut,* 28, 293, 1974.

169. **Wirtschafter, Z. T. and Bichel, M. G.,** Reticuloendothelial response to benzene, *Arch. Pathol.,* 67, 146, 1960.

170. **Goldwater, L. J.,** Disturbances in the blood following exposure to benzol., *J. Lab. Clin. Med.,* 26, 957, 1941.

171. **Snyder, C. A., Goldstein, B. D., Sellakumar, A. R., Wolman, B., Blomberg, I., Erlichman, M. N., and Laskin, S.,** Hematotoxicity of inhaled benzene to Spraque Dawley rats and AKR mice at 300 ppm, *J. Toxicol. Environ. Health,* 4, 605, 1978.

172. **Bernard, J.,** La lymphocytose benzeniques, *Sang,* 15, 501, 1942.

173. **Vianna, N. J. and Polan, A.,** Lymphomas and occupational benzene exposure, *Lancet,* 1, 1349, 1979.

174. **Enderline, P. E.,** Lymphomas and benzene, *Lancet,* 2, 1021, 1979.

175. **Norseth, T., Andersen, A., and Gildvedt, J.,** Cancer incidence in the rubber industry in Norway, *Scand. J. Work Environ. Health,* 2, 69, 1983.

176. **Li, F. P., Fraumani, F., Mantel, N., and Miller, R. W.,** Cancer mortality among chemists, *J. Natl. Cancer Inst.,* 43, 1159, 1969.

177. **Olin, R.,** Leukemia and Hodgkin's disease among Swedish chemistry graduates, *Lancet,* 2, 916, 1976.

178. **Olsson, H. and Brandt, L.,** Occupational handling of chemicals proceeding Hodgkin's disease in man, *Br. Med. J.,* 2, 580, 1979.

179. **Olsson, H. and Brandt, L.,** Occupational exposure to organic solvents and Hodgkin's disease. A casereferrant study, *Scand. J. Work Environ. Health,* 6, 302, 1980.

180. **Torres, A., Grilt, M., Reichs, A.,** Coexistence de antecedentes benzolicas cronicos Y plasmocitoma multiple. Presentation de das cases, *Sangre,* 15, 275, 1970.

181. **Aksoy, M., Erdem, Ş., Dinçol, G., Kutlar, A., Bakioğlu, I., and Hepyüksel, T.,** Clinical observations showing the role of some factors in the etiology of multiple myeloma, *Acta Haematol.,* 71, 116, 1984.

182. **Aksoy, M.,** Lung cancer and chronic benzene poisoning, paper presented at the Proc. Int. Workshop of Toxicology of Benzene, Paris, November 9—11, 1976.

183. **Sherwood, R. J. and Carter, F. W. G.,** The measurement of occupational exposure to benzene vapour, *Ann. Occup. Hyg.,* 13, 125, 1970.

184. **Delzell, E., Anjelkovich, D., and Tyroler, H. A.,** A case-control study of employment experience and lung cancer among rubber workers, *Am. J. Ind. Med.,* 3, 393, 1982.

185. **Snyder, R. and Kocsis, J. J.,** Current concepts of chronic benzene toxicity, *CRC Crit. Rev. Toxicol.,* 3, 265, 1975.

186. **Sims, P., Grover, P. L., Toshio, K., Huberman, E., Marquardt, H., Selkrik, J. J., and Heidelberger, C.,** The metabolism of benzene (a) anthracene and diberiz (a) anthracene and their related K-region epoxides as dihydrolials and phenols by hamster embryo cell., *Biochem. Pharmacol.,* 22, 1, 1973.

187. **Grover, P. L. and Sims, P.,** K region epoxides of polycyclic hydrocarbons reaction with nucleic acids and polynucleotides, *Biochem. Pharmacol.,* 22, 661, 1973.

188. **Jerina, D. M., Daly, J. W., Witkomp, B., Zalterman-Nierenberg, P., and Underfriend, S.,** Role of arene-oxide-oxepin system in the metabolism of aromatic substrates. I. In vitro conversion of benzene oxide to a premercapturic acid and a dihydrodial, *Arch. Biochem. Biophysics.,* 128, 176, 1968.

189. **Jerina, D. M. and Doly, J. W.,** Role of arena oxides: a new aspect of drug metabolism, *Science,* 185, 573, 1974.

190. **Snyder, R., Lee, E. W., Kocsis, J. J., and Witmer, C. M.,** Bone marrow depressant and leukemogenic actions of benzene, *Life Sci.,* 21, 1709, 1977.

191. **Hektoen, L.,** Effect of benzene on production of antibodies, *J. Infect. Dis.,* 19, 69, 1916.

192. **Volkova, Z. A.,** Action of chronic benzene intoxication on phagocytic activity of rabbits, *Gig. Sanit.,* 24, 80, 1959.

193. **White, W. C. and Gammon, A. M.,** The influence of benzol inhalation on experimental pulmonary tuberculosis in rabbits, *Trans. Assoc. Am. Physicians,* 29, 332, 1914; as cited in **Laskin, S. and Goldstein, B. D. Eds.,** Benzene Toxicity: A Critical Evaluation, McGraw-Hill, New York, 1978; *J. Toxicol. Environ. Health Suppl.,* 2, 69, 1977.

194. **Winternitz, M. C. and Hirschfelder, A. D.,** Studies upon experimental pneumonia in rabbits. Paris I and II, *J. Exp. Med.,* 11, 657, 1913.

195. **Bennet, M.,** Effect of age on immune functions on terms of chemically induced cancers, *Environ. Health Perspect.,* 29, 17, 1979.
196. **Heubner, R. and Todero, G. J.,** Oncogenes of RNA tumor viruses as determinants of cancer, *Proc. Natl. Acad. Sci.,* U.S.A., 64, 1087, 1969.
197. **Uyeki, E., Ashkar, A., Shoeman, A., and Bisel, T.,** Acute toxicity of benzene inhalation to hemopoietic precursor cells, *Toxicol. Appl. Pharmacol.,* 40, 49, 1977.
198. **Gill, D. F., Jenkins, V. K., Kempen, R. R., and Ellis, S.,** The importance of pluripotential stem cells in benzene toxicity, *Toxicology,* 16, 163, 1980.
199. **Horigaya, K., Miller, M. E., Cronkite, E. P., and Drew, R. T.,** The detection of in vivo hematotoxicity of benzene by in vitro liquide bone marrow cultures, *Toxicol. Appl. Pharmacol.,* 60, 346, 1981.
200. **Quesenburry, P. and Levitt, L.,** Hematopoietic stem cells, *N. Engl. J. Med.,* 301, 755, 819, 868, 1979.
201. **Stewart, F. M. and Hess, L. E.,** Acute non lymphocytic leukemia and acute aplastic anemia, *Arch. Intern. Med.,* 143, 1156, 1983.
202. **Mallory, T. B., Gall, E. A., and Brickley, W. J.,** Chronic exposure to benzene, *J. Ind. Hyg. Toxicol.,* 21, 355, 1939.

INDEX

L

M

N

T - #0677 - 101024 - C0 - 253/174/9 - PB - 9781138557697 - Gloss Lamination